MATERIAL GEOGRAPHIES OF
HOUSEHOLD SUSTAINABILITY

Material Geographies of Household Sustainability

Edited by

RUTH LANE
Monash University, Australia

and

ANDREW GORMAN-MURRAY
University of Wollongong, Australia

LONDON AND NEW YORK

First published 2011 by Ashgate Publishing

2 Park Square, Milton Park, Abingdon, Oxon OX14 4RN
711 Third Avenue, New York, NY 10017, USA

Routledge is an imprint of the Taylor & Francis Group, an informa business

First issued in paperback 2016

British Library Cataloguing in Publication Data
Material geographies of household sustainability.
   1. Consumption (Economics) 2. Households–Economic aspects. 3. Economics–Sociological aspects. 4. Sustainable development–Government policy.
   I. Lane, Ruth. II. Gorman-Murray, Andrew.
   339.4'7–dc22

Library of Congress Cataloging-in-Publication Data
Lane, Ruth, 1960–
Material geographies of household sustainability / by Ruth Lane and Andrew Gorman-Murray.
      p. cm.
Includes index.
ISBN 978-1-4094-0815-4 (hardback : alk. paper)
1. Sustainable living—Social aspects. 2. Household ecology—Social aspects. 3. Material culture. 4. Environmental protection—Citizen participation. 5. Environmental policy. I. Gorman-Murray, Andrew. II. Title.
   GE196.L36 2010
   577.5'54—dc22

2010047685

ISBN 978-1-4094-0815-4 (hbk)
ISBN 978-1-138-26833-3 (pbk)

Contents

PART III   GOVERNANCE AND CITIZENSHIP

List of Figures and Tables

Figures

Tables

Notes on Contributors

Louise Crabtree is Research Fellow at the Urban Research Centre at the University of Western Sydney. Her research focuses on the social, ecological and economic sustainability of community-driven housing developments in urban Australia, the uptake of housing innovation in practice and policy, and the interfaces between sustainability, property rights and democracy.

Aidan Davison is Senior Lecturer in Human Geography in the School of Geography and Environmental Studies at the University of Tasmania. The author of *Technology and the Contested Meanings of Sustainability* (SUNY 2001), Aidan has published widely on politico-cultural topics related to sustainable development, environmentalism and everyday life.

Robyn Dowling is an urban and cultural geographer. Her primary research interests are in the cultures of home and neighbourhood of suburban residents. Her contribution here draws on an Australian Research Council project concerned with style, fashion, home and family, while she is currently exploring the contours of privatisation and privatism in residential life in Sydney. Her work has been published in the 2006 book *Home* (Routledge, with Alison Blunt, 2006) as well as in *Environment and Planning A*, *Geoforum*, *Political Geography* and *Housing, Theory and Society*.

Chris Gibson is Professor in Human Geography at the University of Wollongong. Recent research projects have explored various aspects of cultural sustainability: in cultural-economic discourse as everyday practice, and as questions for cultural planning. He currently holds an Australian Research Council Future Fellowship exploring the cultural economy of household responses to climate change and financial crisis.

Nick Gill is a human geographer at the University of Wollongong and a member of the Australian Centre for Cultural Environmental Research. He has worked at the National Museum of Australia and the University of New South Wales, where he completed his PhD. He has published on arid zone natural resource management and pastoral culture, Indigenous pastoralism and urban environmental management.

Andrew Gorman-Murray is Research Fellow in Human Geography at the University of Wollongong. His research interests include geographies of home, domesticity and belonging, domestic material culture studies, and geographies of gender and sexuality. He is currently conducting an Australian Research Council funded investigation into shifting spatialities of domesticity and masculinity in inner Sydney.

Gay Hawkins is Professorial Research Fellow and Deputy Director of the Centre for Critical and Cultural Studies at the University of Queensland. In 2006 she published *The Ethics of Waste: How We Relate to Rubbish*, an investigation of the role of waste in the materialisation of environmental ethics (Rowman & Littlefield). She has also published papers on more-than-human politics, assemblage theory, and the relations between environmental publics and markets. She is currently writing a book with Kane Race and Emily Potter called *Plastic Water*, an international investigation of the social and material life of bottled water.

Lesley Head is an Australian Laureate Fellow and Director of the Australian Centre for Cultural Environmental Research at the University of Wollongong. Her research focuses on long-term changes in the Australian landscape and the interactions of peoples with these environments. Her research interests include conceptual debates about culture and nature; sustainability and climate change adaptation; urban natures; Aboriginal land use; and Australian prehistoric environments and human interactions.

Kersty Hobson is Senior Research Fellow at the Environmental Change Institute, University of Oxford. She has held research and teaching posts at the Australian National University and Birmingham University where she has undertaken research into household sustainable consumption, environmental non-government organisations in Singapore and social responses to climate change in the Australian Capital Region.

Ralph Horne is Director of the Centre for Design at RMIT University. He is a researcher of sustainable urban systems, including lifecycle environmental assessment and sustainable design. With a particular interest in housing performance and household social practices, he coordinates research teams and undertakes a variety of projects concerning the environmental, social and policy contexts of production and consumption in the urban environment.

Ruth Lane is Senior Lecturer in Human Dimensions of Environment and Sustainability in the School of Geography and Environmental Science at Monash University in Melbourne. Her research interests are in environmental governance and the social and cultural dimensions of land use change and environmental sustainability. While her current research focuses on issues of consumption and waste in urban contexts, she has also worked on land use and environmental change in rural Australia.

Cecily Maller is a VicHealth Research Practice Fellow at the Centre for Design, RMIT University, Melbourne. Her research is informed by sociology, sustainability, public health and housing studies and focuses on daily routines and interactions between people and natural, built and social environments. Research she is currently pursuing includes exploring the planned, emergent and accidental health outcomes for residents in a master-planned estate, household consumption of energy and water, and adaptation to climate change.

Willem Paling is a final year doctoral candidate at the Centre for Cultural Research at the University of Western Sydney. His doctoral research and recent publications focus on the ways in which modern identities are being forged in Phnom Penh, particularly in the present context in which capital flows of trade and aid, and the cultural flows of tourism and popular culture, have become engaged with Asia.

Emma Power is a lecturer in Geography and Urban Studies at the University of Western Sydney. Her research examines urban natures, the everyday practices of sustainability and homemaking, and human–animal relations. Her PhD, titled *A More-than-Human Geography of Homemaking*, examined the ways in which people interact with nature and non-human animals in the home and garden.

Kane Race is Senior Lecturer in the Department of Gender and Cultural Studies at the University of Sydney. He has published widely in the areas of HIV, sexuality, gay culture and drug use, and is the author of *Pleasure Consuming Medicine: The Queer Politics of Drugs* (Duke University Press 2009). He is currently writing a book with Gay Hawkins and Emily Potter called *Plastic Water*, an international study of the social and material life of bottled water. In other research, he is experimenting with concepts from science studies to explore how consumption practices and their government remake space in a number of different domains.

Andy Scerri is Research Fellow in the Community Sustainability Program of the Global Cities Institute at RMIT University, Melbourne. His research currently focuses on the politics of citizenship and sustainable development in Melbourne and Vancouver, two of 'the world's most liveable cities'.

Gordon Waitt is Associate Professor of Human Geography at the University of Wollongong and a member of the Australian Centre for Cultural Environmental Research. His research interests include urban, rural and nature-based tourism, sustainability and climate change adaptation, and geographies of sexuality and gender. He is currently chief investigator of an Australian Research Council Discovery Grant 'Making Less Space for Carbon: Cultural Research for Climate Change Mitigation and Adaptation'.

Matt Watson is lecturer in Human Geography at the University of Sheffield. His research explores themes of consumption, waste and mobility to advance understandings of the dynamics of everyday practices in relation to sustainability, and the means through which those practices are shaped and governed.

Tim Winter is Senior Research Fellow at the Centre for Cultural Research, University of Western Sydney. He is author of a number of books and articles on heritage, modernity, nationalism and tourism in Asia. His current project, SOCooLH (Sustaining Our Cool Living Heritage), examines the relationship between air-conditioning, modernity and low carbon living in Asia. He recently held visiting scholar positions at Cambridge University and the Getty Conservation Institute.

Acknowledgements

We gratefully acknowledge the support of the Australian Research Council's Cultural Research Network and RMIT University for sponsoring a stimulating research symposium titled 'Material Geographies of Household Sustainability' at RMIT University, Melbourne, in September 2009. The impulse for the project was due in part to an international collaboration with the University of Sheffield, made possible by the British Academy through an Association of Commonwealth Universities grant in 2009. Additional support for the writing and editing phase was provided by the School of Geography and Environmental Science at Monash University and an Australian Research Council Discovery Grant (DP0986666) awarded to Andrew Gorman-Murray.

We would like to thank all those who participated in the symposium for their stimulating commentary and debate and to recognise the contribution of the editors and readers at Ashgate who have assisted in the publication of this collection.

Chapter 1

Introduction

Ruth Lane and Andrew Gorman-Murray

The household, a key focus for many areas of government policy, now looms very large in debates about urban environmental sustainability. The desire for change in household practices informs environmental policy, education and awareness campaigns and is increasingly framed in terms of public ethics and individual responsibilities for environmental sustainability. The geographic scale of the household, we suggest, is more familiar and comprehensible for many lay people than most of the other geographic scales involved in discussions of environmental sustainability, and consequently lends itself to a more inclusive and broad ranging public debate. However for this very reason it also entails many assumptions about the 'normal' practices of mundane domestic life and the motivations of householders for their behaviour and decisions.

The impulse for this collection comes from our recognition of the need to engage more critically with the normative assumptions of household practice within public debates around household sustainability. Sustainability, as it is invoked in this collection, has both material dimensions, in terms of the 'impacts' of a household's resource use, energy consumption, carbon emissions and so on, but also rhetorical dimensions, in terms of public debates about the need to both reduce and adapt to human-induced environmental change. These dual but interrelated concerns require an interdisciplinary approach and we consequently bring together recent work from human geography and cultural studies that focuses on household sustainability from a range of angles, all of which we attempt to capture under the umbrella term 'material geographies'.

Dominant neo-liberal political strategies have emphasised the role of individual 'citizen consumers' in making consumption choices and downplay the more collective forms of social action visible in socio-cultural analysis. We argue that there is a need for research and policy that focuses on householders as stakeholders, and how people influence each other's assumptions about normal and expected behaviours, that is, the *meso* level of social organisation, especially households and face-to-face social networks (Reid, Sutton and Hunter 2010, Fowler and Christakis 2008). This contrasts with social research pitched either at the *macro* scale of the nation, region or population (often resulting in top-down and price-driven policies), or at the *micro* scale of individual psychology (typically relying on theories of reasoned action that do not apply to the unconscious habits and norms of daily living). The household is an organisational unit over which householders have significant control, yet it also articulates with larger scale social processes. While there is great diversity at this scale, there are also observable patterns that are relevant to the spread of pro-environmental and adaptive practices in social networks (Head 2010). As such, interactions at the meso scale of the household provide a network where 'macro level change can be observed and micro level activity can be contextualized' (Reid, Sutton and Hunter 2010: 315).

The challenge and opportunity provided by a material geographies approach to interrogating this meso scale is to consider the operations of the household in terms of interactions between different animate and inanimate entities. We take our cue here from recent assertions (originating in science and technology studies) of the need to move beyond the subject/object or culture/nature dichotomy in order to consider how aspects of the human and non-human world act in conjunction with one another to produce either change or stasis in any given situation. For our purposes therefore, the term 'household' is used to refer to all of these dimensions. For the authors contributing to this collection, the matter involved in a household may include the living and non-living elements of the neighbourhood, the residential block, the dwelling itself, the bodies of those who reside there and the objects, resources and materials that move through the dwelling.

While materiality has long been a concern of geography, especially Marxist geography, the field of 'material geographies' has expanded rapidly over the last 20 years. This is both a consequence of and a response to the cultural turn in human geography. Two key position papers published by Peter Jackson and Chris Philo in 2000 (Jackson 2000, Philo 2000) argued that an over emphasis on the symbolic aspects of culture and on a related politics of individual identity was moving human geography away from concerns about social inequalities and power relations associated with the material realities of people's lives. Both scholars were strongly influenced by the work of Daniel Miller, whose ethnography of shopping drew attention to the significance of ordinary consumption in social relations within households and extended social networks (Miller 1998). Many geographers have shared this concern, but have not endorsed the call to 're-materialise' cultural geography because the notion of materiality employed in these critiques seems to reassert an object–subject dichotomy in a rhetorical strategy that privileges notions of objective reality (Kearnes 2003, Anderson and Wylie 2009). As Noel Castree notes, this dualism is characteristic of post-Enlightenment thinking around culture and nature, and has given rise to a habit of opposing the material and the ideal as 'hard' facticity and 'soft' discursivity (Castree 2003). Our project to develop more critical understandings of household sustainability requires us to explore new approaches to materiality that move beyond this type of dualism.

In a review of recent work on material geographies, Ben Anderson and John Wylie (2009) identify three significant clusters within this literature: material cultures work focusing on 'meaningful practices of use and encounters with objects and environments' (p.318); writing on the intertwined materialities of nature, science, and technology; and materiality around the spatialities of the lived body and its physical and emotional needs, practices and desires. Threads developed through these various bodies of work provide avenues to re-imagine relations between the material, perceptual, emotional and discursive. Key to this agenda is a broader understanding of matter as 'that which is common across other distinctions: between life and nonlife, between the natural and the artificial, and between the organic and the inorganic' (Anderson and Wylie 2009: 319). This inclusive understanding takes in the body and its sensory capacities, manufactured things and all aspects of nature and the non-human world. The locus of the household, or the home, and issues of environmental sustainability are both centrally positioned in these broader debates about materiality.

We will now briefly summarise the three main theoretical developments that inform the clusters of literature identified above and suggest how each may have something to offer for understanding the material dimensions of household practice.

Culture as Agency

Within postmodern accounts of society the emphasis on 'culture' as agency has facilitated greater recognition of the interconnections between economic, cultural and social spheres. It has both broadened out traditional definitions of economic activity and shown how the conventional 'economy' is embedded in social practices, values and habitual behaviour rather than simply material transactions driven by economic motivations alone. For example, how can we understand why householders continue to purchase bottled water when good quality tap water is freely available unless we accept that multiple other values and social practices are at work? This is particularly important as a source of critique for government policy approaches that assume price signals to be the primary motivation for household consumption practices.

Influenced by these developments, geographies of retail and consumption – which previously focused on linear commodity chains linking producers with consumers – now recognise a much broader range of actors and networks involved in production (Cook 2004, Cook and Harrison 2007, Cook et al. 2007), consumption (Cook and Crang 1996, Leslie and Reimer 1999) and, more recently, disposal (Gregson and Crewe 2003, Gregson, Metcalfe and Crewe 2007, Gregson 2007). One example is Alex Hughes and Suzanne Reimer's (2004) notable critique of the global commodity chains literature as being, on the one hand, too focused on systematic links within the chains and, on the other hand, too limited in its understanding of the role of *actual sites* of retail and consumption as producing and impelling practices and meanings of consumption. Recent work has therefore recognised a much wider range of sites of consumption, and has moreover posited these as simultaneously material sites for commodity exchange as well as 'symbolic and metaphoric territories' (Crewe 2000: 227). In the context of this renewed interest in locating and interrogating sites of consumption, the material/symbolic space of the home and household looms very large as a key site in which cultural practices and meanings of consumption are created, embedded and transformed.

Indeed, within the broader field of consumption studies there has been a renewed focus on examining ordinary everyday consumption practices, especially those associated with consumption in the home (Gronow and Warde 2001) and even of the dwelling itself (Bridge 2001). This work also interrogates the cultural meanings of consumption – where consumption is not just a matter of 'exchange values', but also 'use values' and emotional connections – and considers how meanings attached to certain goods are transformed within and across different phases of commodity exchange and circulation by household consumers themselves as critical 'meaning-making agents' (Hughes and Reimer 2004, Cook et al. 2007). A consequence of the new recognition of household consumers as active agents making conscious choices has been new considerations of the ethics of consumption practices and ideas of responsibility and obligation associated with the 'active consumer' or the 'citizen consumer' (Miller and Rose 1997, Barnett et al. 2005, Clarke et al. 2007). Assumptions about consumer responsibilities are now embedded in much government policy around household sustainability, highlighting the need for more critical analysis in this area (Humphery 2010).

In material culture studies object and subject are often posed in a dialectical relationship, drawing on Marxian traditions of thought. In this vein the work of Marcel Mauss in the 1920s on the practice of gift-giving in the Trobriand Islands (Mauss 1969) opened up a key debate within anthropology about the concept of value in society. While some argued that all values are instrumental in some sense, others held that there is a different category of value that cannot be traded along with the possessions that signify it. Some things may be inalienable in the sense that they retain their meaning and identity regardless of who owns or holds them (Levi-Strauss 1969, Sahlins 1972). A particularly productive development of this tradition is found in the work of Arjun Appadurai and Igor Kopytoff on 'the social life of things' (Appadurai 1986, Kopytoff 1986). Their work shows how specific objects may move through 'regimes of value' in the course of their lives as objects. In some phases they may be valued as tradeable commodities, while in others they may be reified as 'inalienable possessions' too precious to be bought or sold (Weiner 1992). This work has much to offer for understanding the meaning of material culture in social systems and is particularly useful for understanding the links between property, status and social agency (Strathern 1999). To date there has been relatively little attention paid to the application of this form of analysis to the more mundane material relations entailed in domestic life, at least in relation to sustainable living.[1] While not generally applied to non-manufactured objects from nature,[2] some geographers now recognise the potential of this approach to inform

---

1   Work by Daniel Miller (2001, 2008) and Nicky Gregson (2007) is arguably influenced by these concepts, but their focus is not on household sustainability.

2   Some work on tourism has shown the potential to apply a similar dialectical analysis to practices of commodification and reification of places and landscapes in the tourism industry: Rojek and Urry (1997), Lane and Waitt (2007).

broader understandings of political ecology around the use of resources such as energy and water in urban environments (Kaika 2004, Bakker and Bridge 2006).

Ideas of Hybridity in Relationships between Nature and Culture

There has been a long-running tension between environmental sciences, which focus on the agency of nature, and the social sciences, which focus on human agency. Prior to the development of more humanistic veins in geography, key texts tended to present human activity as impacting on the natural environment in a simple cause and effect relationship (Johnston and Sidaway 2004). In geography, work by the Berkeley school in the 1950s on cultural landscapes posed culture as the agent and the natural area as the medium (Whatmore 2002). As a result, the traditions of cultural geography that emerged tended to fall into the same trap as other social sciences in privileging human activity over the potential agency of the non-human world. However the development of Actor Network Theory (ANT) in science and technology studies offers a means of redistributing the 'agency' of what might now be regarded as socio-material assemblages. Work by Bruno Latour, Michael Serres, John Law and Michal Callon, on the one hand, and by Donna Harraway and other feminist scholars, on the other, has attempted to break down traditional understandings of subject and object by recognising social agency in non-human entities. This has given rise to what some have termed 'more than human geographies' (Whatmore 2002).

There are key differences between ANT and material culture approaches. While the subject matter of material culture studies is the construction of human social relations, the subject matter of ANT is the construction of knowledge. Material culture studies derive social meaning from the dialectic interplay between (non-human) objects and (human) subjects. Latour's approach to dissolving the object/subject dichotomy involves teasing out the component parts of a socio-material system and treating them all as though they have some form of equivalence as 'actants' working in conjunction with other component parts. In his terms, 'Objects are never assembled together to form some other realm .... Their action is no doubt much more varied, their influence more ubiquitous, their effect much more ambiguous, their presence much more distributed than these narrow repertoires' (Latour 2005: 85). This approach has much to offer for understanding the material dimensions of household practice. Unlike the material culture tradition, allowance is made for agency in nature but, in the context of the household, always in combination with other forms of agency, so that neither the agency of nature nor human agency can be understood to be deterministic in its own right. Work by Emma Power (2007, 2009) and Russell Hitchings (2003, 2004) on living with non-human actants at home demonstrates the keen insights that are gained from this approach.

While there seems to be much potential for ANT to help build new structures of thought that can transcend the nature/culture divide, more detailed well-focused empirical work is needed to unpack the many forms of 'agency' that may be exerted by different sets of entities in specific contexts and to recognise patterns in how they are ordered. Castree offers the following qualification:

> to truly grasp nature's materiality they are going to have to more forcefully explore and acknowledge truly tactile, sensual and embodied ways of figuring the culture-nature nexus. This would entail a relational approach to culture-nature but one where the material capacities are myriad, variable, lively and shared (see Ingold 2000). This approach would be sensitive to historical-geographical difference but would also look for the ways in which relations to nature are nonetheless consequentially ordered in a capitalist world. (Castree 2003: 180)

ANT has been influential in two key areas of study that are particularly relevant to our theme of household sustainability. Firstly, within the sociology of consumption, it has informed a growing literature on socio-technical systems, much of which has specifically focused on the domestic household as an assemblage of materials and practices involving ordinary consumption and habitual behaviour (Shove 2003, Hand and Shove 2007). Alan Warde (2005) argues that the influential theoretical models of consumption proposed by sociologists Zigmunt Bauman and Anthony Giddens rely too heavily on the agency of autonomous consumers preoccupied with symbolic communication. He argues that consumption is not in itself a practice, but an element of a wide range of diverse practices and that these have 'a trajectory or path of dependency' that is conditional on specific institutional arrangements characteristic of particular times, places and social contexts (Warde 2005: 139–40). These might include patterns of household organisation, modes of economic exchange and cultural traditions. Acknowledging a measure of path dependency has implications for understanding the agency of individual consumers, suggesting that we should attend more to the collective development of norms of conduct in everyday life rather than focus on the values and behaviour of individuals (Warde 2005: 149).

Of particular significance is the way in which complexes of material artefacts, conventions and competencies co-evolve within the dynamics of everyday practice (Shove et al. 2007). Shove and co-authors (2007) argue the need to better comprehend the role of materials, tools and technologies in the making, reproduction and transformation of practices in order to understand how 'constellations of products and practices co-evolve' and, in turn, how these constellations relate 'to cycles of production, consumption and innovation' (Shove et al. 2007: 14). These questions around practice highlight other questions about the development of knowledge and competence which can be readily applied to complexes of practice in the household, for instance, do-it-yourself home decorating or renovating (for example, although from a cultural history perspective, Gelber 2000).

The second area of relevant work where ANT has been influential is the rapidly growing literature around environmental governance and citizenship. Here the idea of political agency being distributed among human and non-human entities is gaining traction. Examples of work in this vein that address household sustainability include Éric Darier's study of domestic garbage recycling in Halifax, Canada (Darier 1996), work by Harriet Bulkeley, Matt Watson and others on governance of municipal waste in the United Kingdom (Bulkeley et al. 2005, Bulkeley, Watson and Hudson 2007) and Kersty Hobson's work on the 'techno-ethics of sustainable living' (Hobson 2003, 2004, 2006). This literature draws together aspects of 'network thinking' with Foucault's analytics of government in order to describe and critique some of the shifts occurring in the relationships between government, corporate and community actors around nature conservation, emphasising the roles played by both individual and collective subjects within broader networks of power relations that operate on multiple geographical scales (Bulkeley 2005). Foucault's use of the term 'governance' takes in a much wider cast of actors than formal agents of government, including non-State actors such as environmental NGOs, corporations, community organisations, and the like. Power relations are construed on different but overlapping scales ranging from the individual social identities of householders through to transnational organisations and the operations of international treaties and trade agreements. This represents a significant departure from analyses of power that simply focus on institutions of formal government.

A particular contribution here is the insight provided into the internalisation of power relations by individuals as norms of social conduct (Rose 1999). The discourse and practices promoted by various forms of environmental governance carry new forms of social and environmental ethics and associated responsibilities which, ultimately, influence the way in which individuals understand their own identity in relation to both society and nature. Some scholars have used the term 'environmental citizenship' to describe this phenomenon (Agrawal 2005, Rutherford 2007). For example, concerns about rising levels of atmospheric carbon dioxide and their implications for global climate change give rise to new sets of assumptions about the responsibility of individual householders to reduce energy use in the home.

The term 'governmentality' refers to a broad range of tactics and strategies of government that includes the structures and procedures of institutions as well as the technical means of implementing these procedures. These are concerned with what Foucault termed 'the conduct of conduct' or the establishment and oversight of norms of social practice and behavioral conduct (Rose 1999). In the domestic context these might include government subsidies for water tanks or roofing insulation, as well as restrictions on garden watering. It might also include the forms and procedures that government agencies require households to complete when applying for subsidies for environmental improvements or the presentation of information about resource use on bills from energy or water utilities. A key contribution of Foucault's analytics of government is its central concern with the exercise of power through such procedures and practices. This offers a more sophisticated understanding of the political ecology of household sustainability that highlights the incongruence between the depiction of environmental problems, such as global warming, as commonly shared, and the depiction of their solutions in terms of individual responsibility for consumption practices.

Material Dimensions of Embodied Experience

The final theme of material geographies we wish to highlight is concerned with the 'matter' of our bodies – with corporeality, performativity and embodied experience.[3] While hybrid geographies rework the nature/culture binary, this cluster of work is primarily impelled to disrupt another exemplar of material/immaterial dualisms, that between mind and body. This work, pioneered largely by feminist scholars, focuses on redrawing the notion of 'self' and its relationship to space and society (for example, Longhurst 2001, McDowell 1999, Nast and Pile 1998, Teather 1999). Whereas Enlightenment thinkers and positivist researchers posited 'the self' as an immaterial entity, an essence – or an essentialised and idealised individual mind – feminist geographers reminded us that this 'self' is not ethereal, but is always embodied. This fleshy materiality and its corporeal performances are an indivisible part of our 'selves' and our relationships with other selves and the wider environments in which we live. We 'act' upon this world and upon each other through our bodies, not merely our minds; our conscious and unconscious activities have material consequences for the entwined domains of nature and culture.

For instance, at the most basic level, the corporeal body has material needs for food, clothing and shelter. When these needs are informed (and augmented) by cultural norms around consumption and status, there are implications for environmentally sustainable levels of resource use for home-building, consumer goods, fashion and foodstuffs. While some commentators berate the perceived overconsumption of Western societies (for example, Hamilton and Denniss 2005, James 2007), others offer more creative ways to appeal to positive connections between embodied selves and everyday environments to rethink (and reduce) consumption. In considering how to apply the 'lived and embodied nature of everyday ethics' (Hobson 2008: 201) to encourage more sustainable domestic practices, Kersty Hobson gives the following example:

> [I]nstead of telling us not to drive cars for the obvious reasons – 'peak oil', pollution, obesity, safety, noise – we can appeal to the part of all of us that loves the outdoors, wants interaction and small adventures on the way to work and school, and resents being at the mercy of multinational oil companies and dubious foreign policy decisions. (Hobson 2008: 208)

---

3   Some scholars draw a distinction between 'body' and 'embodiment', suggesting 'embodiment' captures the fluid and processual dimensions of living through the body (for example, McDowell 1999). We agree that embodied experience – including the 'fit' between the physical body and sense of self – is always becoming, but pragmatically we use 'body' and 'embodiment' in an interchangeable manner in this discussion.

Similarly, in examining shifting cultural patterns of domestic water use in western Sydney, Fiona Allon and Zoe Sofoulis (2006: 52) note:

> In addition to demonstrating the complex affective connections between people and their homes and gardens, our research seems to confirm other water-use studies showing a positive correlation between emotional attitudes and feelings towards the garden and the amount of water use outside the home (see Askew and McGuirk 2004). This reveals that the desire to maintain gardens is a strong motivator for people to recycle or collect water.

These provocations are timely: bodies have to consume, but mediating and reducing consumption might best be achieved by underscoring embodied connections to our environments rather than somewhat parenthetical (and parental) hyper-political arguments.

These examples also highlight the spatially relational nature of embodiment: that is, a crucial 'geographical' point of work on embodiment is that we uniquely perform and produce our bodies, materially and discursively, within particular spaces (Duncan 1996, Longhurst 2001, Nast and Pile 1998). Embodied selves and material spaces are mutually constituted in the given context of broader cultural practices. As Derek McCormack, following Judith Butler, argues:

> The 'materialities' of bodies as spaces, and bodies in spaces, can be seen as performative in the sense that their fixities, their contours, their movements are produced through the 'reiterative and citational practices by which discourse produces the effects that it names'. Here the shaping and reshaping of bodies and space are seen as mutually implicated within the other. (McCormack 1999: 158, citing Butler)

As Allon and Sofoulis suggest, one key space where embodied selves are constituted is the home. It is therefore important to consider how geographies of home – especially where they intersect with embodied geographies – can speak to the concerns of household sustainability.

The home is the primary site where embodied demands for shelter, food and clothing are made, and acquiring these material needs is central to the process of 'making home'. Homemaking is thus a performative act which ties the embodied self to a meaningful place, and binds emotional meanings to material possessions and spaces (Blunt and Dowling 2006, Noble 2004). In this process, home and self, and material and emotional connections, become co-constitutive. That is, pre-given individuals do not simply make their homes: rather, we constitute our 'selves' as we build our homes, because this is an environment we care about, one in which we have invested our emotions (Miller 2008). Consequently, there is growing interest in understanding how the material and emotional dimensions of embodied domestic practices have repercussions for environmental sustainability. Kersty Hobson's 2006 work, for instance, examines the embodied techno-ethics of sustainable living at home, exploring how people's adoption of domestic eco-efficient technologies and objects co-constructs their users as 'sustainable citizens'. Here, the body and the home are mutual sites of consumption practices – the consuming body-at-home (Valentine 1999) – and human agency and everyday consumption is, in turn, interdependent on shifting domestic materialities. This embodied nexus of social and technological change can impart positive outcomes for sustainable living but, as Hobson notes, more work needs to be done to see how sustainable practices are enacted in modern eco-homes. Here, cultural approaches help immensely, through ethnographic work, diaries and in-depth interviews.

So far we have moved across linked arguments about the *embodied self* as material and the home as a material *site* for the embodied self. But work on geographies of home shows that the home entails wider material, physical and emotional flows as well. These are arguments about – as Alison Blunt and Robyn Dowling (2006) express it – the home being simultaneously private *and* public. There is widespread popular belief that the modern Western home should be a private space separated from wider social and natural processes (Kaika 2004), sequestered as a site of privatised family life (Blunt and Dowling 2006). But various work has shown home is intimately connected to broader scales through an array of material flows. In her US study of older women with arthritis negotiating their home environments, Pamela Moss argues that the relations that constitute the household are not spatially confined to the physical, material dwelling. They extend outward to automobiles, schools, community organisational meeting places, corner stores and other dwellings. Friendship and social networks also contribute to (re)structuring domestic space through providing emotional support and concrete assistance with maintenance tasks (Moss 1997: 24). Maria Kaika (2004: 274) extends these contentions from social to natural processes (recognising these are entwined), which 'remain connected materially to the inside of the home, constituting an integral part of its material production and its smooth function'. Home-*building* is dependent on consuming natural materials – wood, clay (for bricks), sand (for glass) and ores (for metals). Moreover, homemaking entails ongoing consumption of natural elements – water, energy, air – which 'are *selectively* allowed to enter after having undergone significant material and social transformations, through being produced, purified and commodified' (Kaika 2004: 274). Beyond conscious selection, Emma Power's (2007, 2009) Sydney-based research shows that 'uninvited' nature – primarily non-human animals – also affects homemaking practices, such as patrolling the imagined material borders of the home, with repercussions for embodied feelings of 'unhomeyness' and/or 'homeyness'.

The simultaneous public/private and material/emotional dimensions of home prompt consideration of the politics of homemaking. Just as wider social relationships and material flows intersect in the home, what happens at home can also influence and shape broader social and cultural changes. Scholars have shown the importance of home in effecting wider political changes. Stephen Legg discusses the role of home in the anti-imperial nationalist struggle in Delhi

(1930–1947): 'women helped to politicize the home and assert their agency in a space often read as one of silence and subjection' (Legg 2003: 7) through 'organizing their homes as unofficial political headquarters' (Blunt 2005: 510). Similarly, bell hooks (1991: 42) underscores the radical political dimension of home for African-Americans in the United States, challenging wider processes of racism and sexism by providing a space where they could resist oppression and build personal and collective strength. Equally, as Kersty Hobson (2008) shows, these contentions apply to environmentally sustainable living, or 'green politics' and 'green consumption practices'. While acknowledging the dialectic between government and media messages and behavioural change (Hobson 2002), *enacting* and *embodying* sustainable living nevertheless begins at home: changing domestic consumption practices then flow onto wider social networks, cultural values and further political initiatives (Reid, Sutton and Hunter 2010, Barr and Gilg 2006). Hence, it is important for researchers to understand how sustainable living is engaged at home, and the barriers inhibiting better domestic practices.

Conclusion

The chapters in this volume are organised into three thematic groupings which to some extent align with the approaches and concerns addressed above. The first group consider the contributions of a cultural approach for understanding household sustainability. In particular, the chapters in this section offer a critique of normative assumptions about household practices and behaviour change in relation to a public narrative of environmental sustainability. They also offer a more engaged critical discussion of both assumed and stated conceptions of ethics and responsibilities at the household scale.

The second group considers the patterns and characteristics of domestic spaces and the materials that flow through them. These chapters engage with the material dimensions of homemaking in various forms, from the choice of construction materials, decisions about renovations, to the selection of furnishings for interior spaces and practices around the acquisition and disposal of common household goods.

The third and final group addresses issues of governance and citizenship that emerge in relation to public debates around household sustainability. Neo-liberal governance approaches stress the roles and responsibilities of individual decision makers, however systems of infrastructure and provisioning often establish the parameters for such decisions. The household is a key relational site and scale for mediating these individual and social imperatives (Reid, Sutton and Hunter 2010).

We contend that the material geographies approach presented in this collection goes some way to addressing the problems of unproductive rationalist policy, where '[p]re-determined actions set by national governments and promoted as being 'sustainable' are … constructed in ways that are not reflected in everyday social and environmental concerns' (Barr and Gilg 2006: 907). Ethnographic research is at the forefront of investigating these concerns, while participant action projects can help spread information about sustainability. This form of cultural research, then, might help sustain a productive dialogue between official (government) initiatives for sustainable development, embodied action around the home, and broader ideals of sustainability (Barr and Gilg 2006, Davison 2008).

References

Agrawal, A. 2005. *Environmentality: Technologies of Government and the Making of Subjects*. Durham and London: Duke University Press.

Allon, F. and Sofoulis, Z. 2006. Everyday water: cultures in transition. *Australian Geographer*, 37(1), 45–55.

Anderson, B. and Wylie, J. 2009. On geography and materiality. *Environment and Planning A*, 41(2), 318–335.

Appadurai, A. (ed.) 1986. *The Social Life of Things: Commodities in Cultural Perspective*. Cambridge: Cambridge University Press.

Askew, L. and McGuirk, P. 2004. Watering the suburbs: distinction, conformity and the suburban garden. *Australian Geographer*, 35(1), 17–27.

Bakker, K. and Bridge, G. 2006. Material worlds? Resource geographies and the 'matter of nature'. *Progress in Human Geography*, 30(1), 5–27.

Barnett, C., Cloke, P., Clarke, N. and Malpass, A. 2005. Consuming ethics: articulating the subjects and spaces of ethical consumption. *Antipode*, 37(1), 23–45.

Barr, S. and Gilg, A. 2006. Sustainable lifestyles: framing environmental action in and around the home. *Geoforum*, 37(6), 906–920.

Blunt, A. 2005. Cultural geographies of home. *Progress in Human Geography*, 29(4), 505–515.

Blunt, A. and Dowling, R. 2006. *Home*. London: Routledge.

Bridge, G. 2001. Estate agents as interpreters of economic and cultural capital: the gentrification premium in the Sydney housing market. *International Journal of Urban and Regional Research*, 25, Part 1, 87–101.

Bulkeley, H. 2005. Reconfiguring environmental governance: towards a politics of scales and networks. *Political Geography*, 24(8), 875–902.

Bulkeley, H., Watson, M. and Hudson, R. 2007. Modes of governing municipal waste. *Environment and Planning A*, 39(11), 2733–2753.

Bulkeley, H., Watson, M., Hudson, R. and Weaver, P. 2005. Governing municipal waste: towards a new analytical framework. *Journal of Environmental Policy & Planning*, 7(1), 1–23.

Castree, N. 2003. Geographies of nature in the making, in *Handbook of Cultural Geography*, edited by K. Anderson et al. London: Sage Publications, 168–183.

Clarke, N., Barnett, C., Cloke, P. and Malpass, A. 2007. Globalising the consumer: doing politics in an ethical register. *Political Geography*, 26(3), 231–249.

Cook, I. 2004. Follow the thing: papaya. *Antipode,* 36(4), 642–664.

Cook, I. and Crang, P. 1996. The world on a plate: culinary culture, displacement, and geographical knowledges. *Journal of Material Culture,* 1(2), 131–53.

Cook, I. and Harrison, M. 2007. Follow the thing: 'West Indian hot pepper sauce'. *Space and Culture,* 10(1), 40–63.

Cook, I., Evans, J., Griffiths, H., Morris, B. and Wrathmell, S. 2007. 'It's more than just what it is': defetishising commodities, expanding fields, mobilising change … *Geoforum,* 38(6), 1113–1126.

Crewe, L. 2000. Progress reports, geographies of retailing and consumption. *Progress in Human Geography,* 24, Part 2, 275.

Darier, E. 1996. The politics and power effects of garbage recycling in Halifax, Canada. *Local Environment,* 1(1), 63–86.

Davison, A. 2008. Contesting sustainability in theory-practice: in praise of ambivalence. *Continuum,* 22(2), 191–199.

Duncan, N. 1996. *BodySpace: Destablizing Geographies of Gender and Sexuality.* London: Routledge.

Fowler, J.A. and Christakis, N.A. 2008. Dynamic spread of happiness in a large social network: longitudinal analysis of the Framingham Heart Study social network. *British Medical Journal,* 337: a2338, 23–38.

Gelber, S. 2000. Do-it-yourself: constructing, repairing, and maintaining domestic masculinity, in *The Gender and Consumer Culture Reader,* edited by J. Scanlon. New York: NYU Press, 70–93.

Gregson, N. 2007. *Living with Things: Ridding, Accommodation, Dwelling,* Wantage, UK: Sean Kingston Publishing.

Gregson, N. and Crewe, L. 2003. *Second-Hand Cultures.* Oxford: Berg.

Gregson, N., Metcalfe, A. and Crewe, L. 2007. Moving things along: the conduits and practices of divestment in consumption. *Transactions of the Institute of British Geographers,* 32(2), 187–200.

Gronow, J. and Warde, A. 2001. *Ordinary Consumption.* New York: Routledge.

Hamilton, C. and Denniss, R. 2005. *Affluenza: When Too Much is Never Enough.* Crows Nest: Allen and Unwin.

Hand, M. and Shove, E. 2007. Condensing practices: ways of living with a freezer. *Journal of Consumer Culture,* 7(1), 79–104.

Head, L. 2010. Cultural ecology: adaptation – retrofitting a concept? *Progress in Human Geography,* 34(2), 234–242

Hitchings, R. 2003. People, plants and performance: on actor network theory and the material pleasures of the private garden. *Social and Cultural Geography,* 4 (1) 1, 99–113.

Hitchings, R. 2004. At home with someone nonhuman. *Home Cultures,* 1(2), 169–186.

Hobson, K. 2002. Competing discourses of sustainable consumption: does the 'rationalisation of lifestyles' make sense? *Environmental Politics,* 11(2), 95–120.

Hobson, K. 2003. Thinking habits into action: the role of knowledge and process in questioning household consumption practices. *Local Environment*, 8(1), 95–112.

Hobson, K. 2004. Sustainable consumption in the United Kingdom: the 'responsible' consumer and government at 'arm's length'. *Journal of Environment and Development*, 13(2), 121–140.

Hobson, K. 2006. Bins, bulbs, and shower timers: on the 'techno-ethics' of sustainable living. *Ethics, Place and Environment*, 9(3), 317–336.

Hobson, K. 2008. Reasons to be cheerful: thinking sustainably in a (climate) changing world. *Geography Compass*, 2(1), 199–214.

hooks, b. 1991. *Yearning: Race, Gender and Cultural Politics*. Boston: South End Press.

Hughes, A. and Reimer, S. 2004. *Geographies of Commodity Chains*. New York: Routledge.

Humphery, K. 2010. *Excess: Anti-Consumerism in the West*. Cambridge: Polity Press.

Ingold, T. 2000. *The Perception of the Environment*. London: Routledge.

Jackson, P. 2000. Rematerializing social and cultural geography. *Social and Cultural Geography*, 1(1), 9–14.

James, O. 2007. *Affluenza*. London: Vermilion.

Johnston, R. and Sidaway, J. 2004. *Geography and Geographers: Anglo-American Human Geography Since 1945*. 6th Edition. London: Arnold.

Kaika, M. 2004. Interrogating the geographies of the familiar: domesticating nature and constructing the autonomy of the modern home. *International Journal of Urban and Regional Research*, 28(2), 265–286.

Kearnes, M.B. 2003. Geographies that matter–the rhetorical deployment of physicality? *Social and Cultural Geography*, 4(2), 139–152.

Kopytoff, I. 1986. The cultural biography of things: commoditization as process, in *The Social Life of Things: Commodities in Cultural Perspective*, edited by A. Appadurai. Cambridge: Cambridge University Press, 64–95.

Lane, R. and Waitt, G. 2007. Inalienable places: self-drive tourists in northwest Australia. *Annals of Tourism Research*, 34(1), 105–121.

Latour, B. 2005. *Reassembling the Social: An Introduction to Actor-Network-Theory*. Oxford: Oxford University Press.

Legg, S. 2003. Gendered politics and nationalised homes: women and the anti-colonial struggle in Delhi, 1930-47. *Gender, Place and Culture*, 10, Part 1, 7–28.

Leslie, D. and Reimer, S. 1999. Spatializing commodity chains. *Progress in Human Geography*, 23(3), 401–420.

Levi-Strauss, C. 1969. *The Elementary Structures of Kinship*. Boston: Beacon Press.

Longhurst, R. 2001. *Bodies: Exploring Fluid Boundaries*, London: Routledge.

McCormack, D. 1999. Body shopping: reconfiguring geographies of fitness. *Gender, Place and Culture*, 6(2), 155–177.

McDowell, L. 1999. *Gender, Identity and Place: Understanding Feminist Geographies*. Cambridge: Polity.

Mauss, M. 1969. *The Gift: Forms and Functions of Exchange in Archaic Societies*. London: Cohen and West.

Miller, D. 1998. *A Theory of Shopping*. Cambridge: Polity.

Miller, D. 2008. *The Comfort of Things*. Cambridge: Polity.

Miller, D. (ed.) 2001. *Home Possessions: Material Culture behind Closed Doors*. Oxford: Berg.

Miller, D. and Rose, N. 1997. Mobilising the consumer: assembling the subject of consumption. *Theory Culture and Society*, 14(1), 1–36.

Moss, P. 1997. Negotiating space in home environments: older women living with arthritis. *Social Science and Medicine*, 45(1), 23–33.

Nast, H. and Pile, S. 1998. *Places Through the Body*. London: Routledge.

Noble, G. 2004. Accumulating being. *International Journal of Cultural Studies*, 7(2), 233–256.

Philo, C. 2000. More words, more worlds: reflections on the 'cultural turn' and human geography, in *Cultural Turns, Geographical Turns: Perspectives on Cultural Geography*, edited by I. Cook et al. Harlow, England and New York: Prentice Hall, 26–53.

Power, E. 2007. Pests and home-making: depictions of pests in homemaker magazines. *Home Cultures*, 4(3), 213–236.

Power, E. 2009. Border-processes and homemaking: encounters with possums in suburban Australian homes. *Cultural Geographies*, 16(1), 29–54.

Reid, L., Sutton, P. and Hunter, C. 2010. Theorizing the meso level: the household as a crucible of pro-environmental behaviour. *Progress in Human Geography*, 34(3), 309–327.

Rojek, C. and Urry, J. 1997. *Touring Cultures: Transformations of Travel and Theory*. London: Routledge.

Rose, N.S. 1999. *Powers of Freedom: Reframing Political Thought*. Cambridge and New York: Cambridge University Press.

Rutherford, S. 2007. Green governmentality: insights and opportunities in the study of nature's rule. *Progress in Human Geography*, 31(3), 291–307.

Sahlins, M. 1972. *Stone Age Economics*. Chicago: Aldine-Atherton.

Shove, E. 2003. *Comfort, Cleanliness and Convenience: The Social Organisation of Normality*. Oxford: Berg.

Shove, E., Watson, M., Hand, M. and Ingram, J. 2007. *The Design of Everyday Life*. Oxford: Berg.

Strathern, M. 1999. *Property, Substance and Effect*. London: The Athlone Press.

Teather, E.K. (ed.) 1999. *Embodied Geographies: Spaces, Bodies and Rites of Passage*. London: Routledge.

Valentine, G. 1999. A corporeal geography of consumption. *Environment and Planning D*, 17(3), 329–351.

Warde, A. 2005. Consumption and theories of practice. *Journal of Consumer Culture*, 5(2), 131–153.

Weiner, A.B. 1992. *Inalienable Possessions: The Paradox of Keeping-While-Giving*. Berkeley: University of California Press.

Whatmore, S. 2002. *Hybrid Geographies: Natures Cultures Spaces*. London: Sage Publications.

PART I
Contributions of a
Cultural Approach to
Household Sustainability

Chapter 2

Is It Easy Being Green? On the Dilemmas of Material Cultures of Household Sustainability

Chris Gibson, Gordon Waitt,
Lesley Head and Nick Gill

Introduction

In the 1970s 'greens' were normally thought of as radicals because of their uncompromising political views about sustainability, non-violence, social justice and grassroots democracy. Sometimes greens were marginalised as 'tree-huggers' because of their affinity with the non-human world. Today, in popular discourse, 'green' provides the centre of sustainability gravity (Barr 2003). Green has become a definitive reflection of what individuals are to become as both consumers and citizens. It is easy, it is said, to be green. This is evident from product branding to categories used in government survey results to describe the 'most acceptable' household practices. But as green is drawn into the mainstream of politics and commerce, there are both possibilities and problems. The possibilities are self-evident. The more one might be defined by a 'green identity', the more likely everyday decisions will be informed by choices, attitudes and practices that purport to be 'environmentally friendly'. Individuals start to reflect on everyday practices. Is the tap running as I clean my teeth? Am I carrying my shopping home in a plastic bag? Can I take public transport? But the problems are just as evident. The more green life becomes mainstreamed, the more it shares with consumerism, becoming trendy among the young and affluent and lacking creativity as certain personal actions and consumer products become coded as the 'right' green choices.

We think this path towards redressing human-induced climate change – focusing on 'greening' the behaviour of households vis-à-vis material forms of consumption – is problematic. At the very least assumptions ought to be examined about the supposed ease with which households are to transition towards a set of 'more green' behaviours. As Mike Hulme (2008: 5) has argued, this kind of 'constructed policy goal is unlikely to be one around which the world will be re-engineered willingly'. But also, it seems, assumptions about being green in one's personal conduct and consumption behaviour have quickly become normative, with uncertain potential consequences. Among the risks is that a set of class- and culture-specific practices become entrenched as 'green'.

Our desire to challenge assumptions about what might constitute 'green' household behaviour and the ease with which it is embraced, is borne not of scepticism about climate change, nor from any hesitation about the need to act urgently. Indeed, like Paul Chatterton (2010) and many others, we agree that something radical needs to be done about how we, as humans, use fossil fuels as the basis of an urban industrial capitalist system that for human advantage generates torrents of resource extraction, production and consumption.

Our concerns with the presumption that households can readily make a difference by being more green are instead about dilemmas of practice and circumstance: firstly, that being green may not in fact be so easy or its choices self-evident; secondly, that an emphasis on sustainability in terms of green consumption masks a necessary critique of the industrial capitalist imperative to produce ever more stuff to consume; and thirdly, that embedded in some green sustainability scripts is a placelessness that belies how geographical context and social and cultural meanings vary, thus altering the entire landscape of what sustainability might mean for households. In our work the spatial context we are addressing is both domestic space (referred to as the meso level by Louise Reid, Philip Sutton and Colin Hunter 2009) and the surrounding city. Their relationality to each other and to wider spatial contexts is a dynamic process demonstrating both continuity and change.

We work as a team across a number of interrelated projects on households, sustainability and cultural practices in Wollongong, a small industrial city 80 kilometres to the south of Sydney. Here, understanding and acting on the imperative to be sustainable is refracted through a number of contingencies: Wollongong's industrial heritage (steel, coal) means that it has been 'carbon central'; that these industries are still active means that a substantial proportion of the city's population are not only employed in industries that may be affected by future legislative attempts to reduce national carbon emissions, but that also, within households in Wollongong, individuals are also *producers* of carbon emissions in an industrial sense (Gibson et al. 2011). This creates a rather objectionable situation where federal and state governments are encouraging households to 'be more green' in their consumption behaviour, amidst policy shifts that are believed by some to threaten the region's industries, and thus the livelihoods of those same households trying to do the right thing by the environment in their consumption behaviour. Wollongong's history of being a city exposed to fluctuations of global commodity and manufacturing markets means that there are also territorial loyalties and suspicions here that lurk beneath sustainability discourses. The imperative to 'be greener' has for many Wollongong households become framed as yet another example of local people being penalised by state and national governments and global political manoeuvrings.

But the regional setting complicates matters further: for instance, sceptical framings of climate change have been almost absent from the pages of the city's tabloid newspaper, the *Illawarra Mercury*.[1] Instead, in the context of discussions about a carbon pollution trading scheme, climate change has been framed locally through intersecting discourses of green governmentality, civic environmentalism and ecological modernisation. Readers of the *Illawarra Mercury* are presented with an editorial perspective that accepts the science of climate change, but that mutates discourses about what should be done about it: readers have been 'sold' the need for climate change action because of the impending threat of floods, sea-level rise and bushfires locally, rather than appealing to any altruistic desire to 'save the planet' (that generic prompt assumed to motivate households in the normative 'green' sustainability script). In turn, sustainability in Wollongong is most commonly presented in local media as an agenda beginning in the home rather than at the city's steelworks, for fear of criticising Big Industry in Wollongong (see Hinchliffe 1996). Another influence is the city's unique material geography – a coastal port city hemmed into a narrow coastal strip due to a dramatic, heavily forested escarpment, with 60 kilometres of high amenity surfing beaches. Local concerns about climate change are rendered tangible by newspaper front pages with headlines about job losses at the steelworks resulting from national policy shifts to reduce carbon emissions, or showing maps of which beachside suburbs will lose land and houses in the event of a rise in sea level. The result is a regional context where the binaries that are assumed to characterise the climate change debate globally (sceptics versus 'believers') simply do not hold true, with a far more contradictory mash of competing discourses at play.

This, then, has been the setting for our various research projects seeking to explore household cultures of sustainability in Wollongong.

With these regional idiosyncrasies in mind, in the remainder of this chapter we discuss some of the *dilemmas of household sustainability* as observed in our research in Wollongong (Gibson et al. 2011). Ours is not a hostile response to those who might wish to promote sustainability, but rather a critique of how the purpose of 'green' action is pegged to an all too easy script of transformation about education, altruism, motivation and behaviour. Sustainability has long been recognised as having multiple and contested meanings. Reduced greenhouse gas emissions to mitigate climate change can be understood as a recent subset of wider sustainability concerns. Some of the dilemmas occur when 'sustainability' behaviours are at cross purposes, for example when rainwater tanks to save mains water require installation of an electric pump to bring the water into the house. But we also want to identify the complexities of incorporating normative behaviours into everyday life.

An increasingly normative script – evidenced, for instance, in the New South Wales government's advertising campaigns on billboards and television about reducing carbon emissions by lowering the temperature of air-conditioners – suggests that transformation can be realised by educating people on the need to

---

1   In preparation: Waitt, G., Farbotko, C. and Criddle, B. Imagining climate change regionally (1997–2008).

respond to the challenge of climate change. It appeals to altruistic feelings about doing the right thing by the environment, and assumes that this single motivation will henceforth inform motivations to behave (that is, consume) in ways that are 'green' (pro-environmental) rather than damaging. What our research in Wollongong has found is that things are not nearly so linear. Rather, questions of household sustainability need to be understood in terms of contradiction and ambivalence (Hobson 2003, Davison 2008). The key is to understand the dilemmas of sustainability for households as emergent in the material circumstances of everyday life – how such dilemmas surface and escape resolution. Below we structure our discussion accordingly, around a taxonomy of dilemmas: how to account for the carbon implications of different forms of material consumption in the household; the moralities of material sustainability; and the household in its social context.

To clear space for such a discussion, we have shorn from this chapter the question of how to theorise 'the household'. Suffice to say that the theoretical perspective we bring to this chapter is one informed by the material turn in cultural geography (one increasingly concerned with how bodies, ideologies, technologies and materials are involved in the co-constitution of space) and more specifically by research into the geographies of home (Blunt 2005, Blunt and Dowling 2006, Gorman-Murray 2006, Gregson and Beale 2004, Power 2009) and the situated entanglements of household sustainability (Hobson 2006, Lane, Horne and Bicknell 2009). Such research materialises sustainability in the cultural contexts of home, and conceptualises household practices as spatial expressions of identity work and shifting meanings of domesticity. It also confirms the porosity of home spaces: the various ways of maintaining, or breaking, linkages outwards into wider social, technological and regulatory networks that comprise suburbs, cities, regions and nations. From this perspective, households are not detached units but rather situated in contexts, relationships, 'enrolled networks' and processes of all sorts that guide normative behaviour (Gibson et al. 2011).

Also ejected from this chapter in the interests of brevity are lengthy explanations of the methodologies used in our research projects. At various points we draw selectively from a baseline survey of households – entitled 'Tough Times? Green Times? A survey of the issues important to households in the Illawarra' (henceforth 'Tough Times? Green Times?') – which was compiled using various quantitative measures to support a key flagship project – 'Making Less Space for Carbon: Cultural Research for Climate Change Mitigation and Adaptation' ('Making Less Space for Carbon'), a project funded through the Australian Research Council's Discovery Projects scheme. The baseline survey was the project's first key activity, involving postal distribution in July 2009 of the hard copy survey to 11,555 households, in eight suburbs in metropolitan Wollongong (stratified to account for distribution across quintile income categories). Of this number, 1,465 surveys were completed – a 12.67 per cent response rate. Statistical tests indicated that there was *no* significant difference between the sample population and the total metropolitan Wollongong population as enumerated in the 2006 Australian

Census by household composition, number of people in households, total weekly household income, dwelling type, age, education and employment. It can thus be relied upon as a statistically representative sample for the City of Wollongong. In this chapter we frequently report on results of the 'Tough Times? Green Times?' survey to illustrate our contentions about the dilemmas of sustainability.[2] Orbiting around the 'Making Less Space for Carbon' project have been a constellation of other projects, some large (such as PhD research topics and longitudinal ethnographic work undertaken by ourselves and post-doctoral fellows); others such as honours research, which has been more contained. We occasionally also draw on key insights from these projects in this chapter.

Accounting for Material Sustainability in the Household

Straightforward dilemmas arise from our concern that it seems possible to embrace green rhetoric and craft a 'green identity' through purporting to have a green attitude, being knowledgeable about the 'environment' and purchasing environmentally-friendly products –without making significant reductions in the consumption of resources. At the heart of this are tensions between awareness of what constitutes 'green' consumption and actual measurements of energy and water use tied to instances of consumption – as well as further contradictions between certain practices at the level of individual households.

From statistical modelling in our 'Tough Times? Green Times?' survey it was clear that households already committed to behaviours normatively understood as pro-environmental (recycling, composting) were more likely to express an interest in climate change, a preparedness to change household behaviours, as well as optimism for mitigating greenhouse gas emissions and reducing the potential risks of bush fires and rising sea levels. In comparison, the households categorised as 'reluctant' by their collective environmental behaviours were more likely to express a lack of interest in climate change, and were more fatalistic and fearful about the future implications of greenhouse gas emissions. This was not entirely unexpected.

However results of 'Tough Times? Green Times?' also suggested that income, household consumption, education and length of residence were not consistent in differentiating between households actively engaged in normatively constituted 'pro-environmental' behaviours. In cluster analysis of strength of commitment to everyday practices of pro-environmental behaviour, we were struck by how households with 'strong', 'modest' and 'limited' commitments were represented across diverse social groupings and educational levels. And while some pro-environmental practices had become habitual for most households across socio-

---

2   For a more comprehensive explanation of method, and statistical analysis of this data see, in preparation, Waitt, G., Caputi, P., Head, L., Gibson, C. and Gill, N. Tackling climate change through reduced consumption? A two-stage cluster analysis of household consumption dynamics.

economic groups (such as recycling, using 'green bags' rather than disposable plastic bags, turning off taps and lights, putting on an extra layer of clothing before turning up the heating and donating used clothing to charities), even the majority of strongly committed households did not report regularly walking to the shops, growing their own fruit and vegetables or purchasing organic food, fair-trade products or recycled toilet paper. There therefore appear to be limits within which even strongly committed households are either able or prepared to embrace practices thought of as pro-environmental.

There were ambiguities too in patterns of consumption outside the realm of pro-environmental behaviour, around air-conditioner use and ownership of fridges and clothes driers. The poorest households (in income terms) had the fewest fridges (74 per cent of households in the lowest income bracket had only one fridge, compared with 33 per cent in the highest income bracket); and household and family types normally understood as socio-economically vulnerable, such as single parent families and retirees living alone, who also had fewer fridges (75 per cent and 85 per cent, respectively, had only one fridge). By contrast, in the highest income bracket the norm was to own two fridges (49 per cent), or even three or four (25 per cent of highest income households). But even this seemingly direct correlation was made more complex when compared against other variables: those holding bachelor or honours degrees from universities were as likely to own two, three or four fridges as those with basic Year 10 high school as their highest level of educational attainment; 43 per cent of two-person households felt the need to own two fridges, while 36 per cent of six-person households managed to get by with just one.

Beyond fridges, the contradictions expanded exponentially: the wealthiest bracket of households were twice as likely to install solar power (although still in very small numbers) as the poorest, but were also the most prevalent users of air-conditioning. The poorest households were most likely to say that they were 'uninterested' in climate change as an issue (22 per cent of the lowest income bracket, compared with 3.6 per cent of the highest income bracket), but they were also the least likely to own LCD or plasma screen televisions or clothes dryers. The poorest households were also the most likely to repair clothing, to use toilet paper made from recycled paper; to buy 'environmentally-friendly' detergents; to reuse glass bottles and jars; and to save water by taking shorter showers (indeed, by a huge margin: 44 per cent of the lowest income bracket 'always' took shorter showers, compared with 5 per cent and 15 per cent of the two top highest income brackets). Those with higher levels of educational qualifications (bachelor, honours and postgraduate university degrees) were equally likely to use air-conditioning habitually during summer as were those with basic high school education (Year 10 or below), but current students, as well as the unemployed, were the least likely to use it. Baby boomers were the least likely to be sceptical about climate change, but the most likely to fly five times or more in a 12 month period. And so on and so on, across a myriad of types of consumption within households from kangaroo meat eating to energy-saving light globes.

All this is then rendered more complex again if we question the actual emissions embedded in different products and the processes connecting them to the household. Does the travel impact of items sold on eBay™ outweigh the recycling value? Do the greenhouse gas emissions used in the production of certain types and scales of food far outweigh transport emissions, as some New Zealand-based work on milk has suggested (Saunders, Barber and Taylor 2006), turning the injunction to 'eat local' into a misguided if not dangerous assumption? (See also Johnston, Biro and MacKendrick. 2009). Recent energy use studies by Choice, the Australian consumer group, demonstrated that even in one electronic appliance – digital set-top boxes – standby and on-power energy use can vary as much as 400 per cent across different brands, all within the same price bracket (Choice Australia 2009).

Householders' concern to do the right thing vis-à-vis consumption of material goods is undercut by obfuscation of their actual environmental impacts. Clothing is a prime example. Chris Gibson and Elyse Stanes (2010) tracked the emergence of 'green fashion' as a new trend in clothing marketing, exemplified by the incorporation of organic cotton into mainstream production of jeans (Levis) and T-shirts (Marks & Spencers). But what isn't clear to consumers is the sheer amount of energy consumed during cotton processing, ginning, balling and dyeing, or the geographical variability of cotton production (in Israel, for instance, cotton production requires 7,000 litres of water per kilogram, while in Sudan the figure is more like 29,000 litres; Fletcher 2008).

The picture is complicated again when material goods whose environmental consequences have already been obfuscated are inserted into the material-cultural geographical spaces of the household. Again, clothing provides a stark illustration. Here meeting the essential human need to be covered intersects with fashion as a *cultural* industry, promoting particular forms of consumption as social practices, where fashion is central to the production of self-identities and a way of marking the body with meaning. Most people own more clothing than is necessary, replace perfectly useable items with changes in season and trends, and wash clothing more often than is needed. In this context fashion branded as 'eco' or 'sustainable' becomes a means to accumulate cultural capital; a new moral economy 'ultimately constrained by the size of niche markets' (Bassett 2010: 2, see also Beard 2008) – while leaving intact the holy grail of more sales and consumption. And there are yet more contradictions about the ways people relate to clothing as fashion. Undertaking ethnographic work on the cultural economy of clothing consumption of young people in Wollongong, Gibson and Stanes (2010) found that those most likely to consume far more clothing than they needed were also the most likely to consider environmental sustainability when making purchases (they were most likely to buy organic cotton jeans or other items of clothing marketed as more sustainable). By contrast, those least likely to buy large amounts of clothing (people describing themselves as frugal or even 'anti-fashion') frequently bought clothing that was cheap, made of poor quality textiles and that carried a heavy environmental burden. When all of these kinds of contradictions and complexities are factored in, exactly where households sit on a balance sheet of accounting for environmental impact becomes decidedly unclear.

Moralities of Material Sustainability

Efforts such as improving sustainable transport and encouraging energy-saving consumption in the home stem from examining what kinds of activities produce bad climate change outcomes. But the theory behind public education campaigns is not particularly responsive when it comes to understanding underlying human behaviour, needs or desires (Malpass et al. 2007). Understanding the household as a site of homemaking practices and a nexus of intersecting ideological and technological relations points to questions of what ethical worlds inform everyday practices, and precisely where altruism comes into play – if at all. What is critical here is that morality emerges from within social practice, not simply out of an aspiration towards abstract altruistic ideals (Gough 2010). The environment is just one line of responsibility being juggled in acts of consumption, which necessarily serve different anticipated needs (Dowling 2000, Burgess et al. 2003). As Kersty Hobson (2006) demonstrates, situation-specific ethical moments come into play. Urging householders to behave responsibly in these moments has its limits, with what constitutes 'responsible' behaviour being developed within social worlds such as households, rather than abstractly distant from them. Hence morality is informed by a sense of social justice; of doing right by one's family, neighbours, friends; by domestic moralities (how best to bring up kids or to ensure their safety); and also by norms of cleanliness, comfort, convenience and waste (Shove 2003, Hawkins 2006) – all of this occurring in the shifting spaces and moments of personal decision making within households.

Financial imperatives and changes in household structure appear to both override and intersect with concerns about the carbon implications of behaviour. In 'Tough Times? Green Times?', when asked for the primary reason for any change in energy consumption or car use in the past 12 months, financial imperatives and changes in household structure always outnumbered explanations based on responding to the problem of climate change. For instance, of those who said that they had experienced a change in personal car use over the previous 12 months, only six per cent registered climate change as a factor, compared with 41 per cent citing being made redundant and 19 per cent having recently retired. This is not to say that environmental ethics are not present: where respondents recorded a change in household energy use, the most commonly cited reasons did include 'purchasing energy-efficient appliances', 'purchasing green power' and 'concern about climate change'; but these were, statistically speaking, *mixed up* among other reasons cited, including increased cost of power; a change in the number of people in the household; recently having a baby; becoming unemployed; or retiring from work.

It must be emphasised too that social practices, cultural norms and the material spaces of the home are iteratively linked: none come pre-formed or absolute. The techniques of design matter, for instance, because they relate to (and are in turn influenced by) shifting everyday practices: some designs for appliances (such as standby circuits on DVD players), furniture and home interiors 'ratchet' up consumption in ways that are difficult to reverse (Shove 2003, Shove et al. 2007); other forms of design create more malleable objects that can be put to different purposes, extending the life of products or orientating them to new uses. An example of where these dilemmas can be thought through is that of big houses, recently built in master-planned estates. Such houses (and estates) are often condemned for their ideological imprint of mass consumption ('McMansions'), but are also valued by families for their sense of open space and flow (see Chapter 5 in this volume). It is true that most houses in new estates have been designed with no eaves to maximise indoor house space on the available land, which means high energy bills for both air-conditioning in summer and heating in winter (through elimination of the 'passive' design quality of eaves allowing or deflecting sunlight differentially in opposite seasons). Although per capita floor space has increased dramatically in recent years, those same houses do have ample space to include more people if future family size and living patterns shift (potentially reducing the overall per capita footprint); and they do enable families to hoard things for future reuse. Indeed we would do well to debate further the politics of hoarding (Gregson and Beale 2004, Lane, Horne and Bicknell 2009): is hoarding an outcome of overconsumption, or a practice that reduces waste disposal and enables future reuse?

Larger houses may be used to enable extended families (grandparents and couple families with children) to co-habit (Borger 2010). Sometimes co-habitation involves altering the footprint of buildings, converting garages into granny flats or expanding homes through extensions and extra floors – adaptations that enable co-habiting families to live 'under the one roof', but as separate families: 'living together but apart'. These arrangements, predicated on living in large houses, at times enable the sharing of resources (such as food) and responsibilities (such as child care) in ways that are likely to reduce per capita consumption.

By comparison, small houses and apartments have their contradictions: they encourage careful purchasing of items for fear of crowding, but also force people to throw out goods more frequently, exacerbating the voluminous throughput of consumables. This is especially so in cities, where small footprint homes (especially apartments) are linked to the gradual downward shift in average family size – itself a problem for sustainability because of the resultant replication of household infrastructure and energy use across a larger number of household units (Lenzen, Dey and Foran 2004).

The use of washing machines, as evidenced in 'Tough Times? Green Times?', demonstrates some of this complexity. Frequency of use appeared consistent across households of all *sizes* (that is, there was no noticeable efficiency gain for larger or smaller households, in terms of numbers of people), but it did vary by household *type*, with extended family households using the washing machine about as often as couple families with children. This suggests that extended families are doing 'communal' clothes washes as a matter of course, consequently meeting the laundry needs of two (or more) co-habiting families with about the same number of washes. Other family types used the washing machine differently, for instance each family member in a nuclear family running a separate wash – with resultant impacts on water use. For dishwashers the story was similar: one person households were least likely to own a dishwasher, or use one daily if they did (washing by hand or saving up dirty dishes to wash every couple of days), while extended families used a dishwasher slightly less often than couple families with children (48 per cent and 55 per cent, respectively, used a dishwasher most days or daily). In other words, as the structure of households and material spaces of co-habitation change so too do the ethics and pragmatics of social worlds begin to shape decisions about resource use, sharing and consumption.

The manner in which the morality of household consumption is shaped by industry logics is also relevant here: in the case of gardening practices such as lawn maintenance and watering, by the petrochemicals industry and privatised utility supply companies (Askew and McGuirk 2004, Robbins 2007); in the case of household cleaning products and behaviours, the petrochemicals industry (again) and the marketing of disinfectant products appealing to fear of germs and post-Second World War norms of cleanliness. In clothing, the pervasiveness of an industry-generated fashion cycle has meant that enormous amounts of clothing production and consumption occur on a seasonal basis, a product of conspicuous consumption rather than basic need (Gibson and Stanes 2010). The fashion cycle has sped up, too: in response to losing manufacturing to China, the US textiles industry deliberately restructured production systems to enable what is known as 'fast fashion' (Doeringer and Crean 2006), whereby localised US manufacturers respond to street fashions practically instantaneously (rather than designing the traditional twice-yearly 'new season' lines). Shops can now change product lines every month if they so desire, thus maximising sales as consumers are tempted by more frequent updates in item availability. The result is a higher overall consumption of clothing and the incorporation of larger wardrobes into house design, which in turn establishes the conditions for further accumulation of clothing (Gregson and Beale 2004). Accordingly, Clive Hamilton and colleagues (2005) estimate that approximately A$1.7 billion dollars is spent annually in Australia on clothing that is not worn.

A spotlight also needs to be cast on the dispersal of fashion cycle logics – originating in the clothing industry – into industries where they once did not apply. Scott Lash and John Urry (1994) described this as the symbolic economy; an enculturation of all forms of production with cultural logics of taste and fashion signification. Computers that were once drab beige or grey boxes are now designer products (thanks to Apple) to be replaced regularly as colours and features change. Home interiors need replacing not because furniture has been broken or because paint needs refreshing, but because art deco 'is in', or fabrics and colours go in and out of style. That this is so is more than mere individual greed or vanity; it is entirely to be expected in a world where cultural capital is a measure of self-worth and where awareness of how others perceive you is paramount.

Amid this swirling mix of industry logics, societal norms and ethical responsibilities to the environment, family, society and self are the lived moments of decision making about everyday practices around the home. Trade-offs between different goods/evils become dilemmas: is it worse to waste the water to rinse out tin cans than to put them in the recycling bin dirty? Is it worse to use plastic supermarket bags for bin liners, or to take reusable green bags to the supermarket but then buy dedicated bin liners? Beyond the need for better calculations of these sorts of trade-offs are dilemmas of time management and everyday household practice. How much time do well-intentioned people spend thinking these choices through or debating them within a household? In our survey 30 per cent of respondents said they felt rushed or pressured for time 'frequently' or 'always' and a further 40 per cent felt rushed or pressured for time 'sometimes'. Less than 5 per cent 'never' felt rushed or pressured for time and only about the same percentage said that they 'frequently' or 'always' had spare time that they didn't know what to do with (contrasting with 30 per cent who 'never' had spare time). We asked our survey respondents to record any arguments they had in their household about recycling, reusing or reducing their environmental footprint – expecting them to be prevalent. But only just over 10 per cent admitted that arguments occurred, evidence perhaps that households simply do not have the time to undertake the required 'thinking work' around sustainability. Beyond the scope of our survey are further questions about which members of households undertake this 'thinking work': who feels guilty; is it worth it, and what could be the outcomes if the same amount of effort was invested elsewhere?[3]

The Household in Social Context: What are the Carbon Implications of Social Processes and Trends?

A third dilemma of sustainability stems from the social actions of households, rather than from material products consumed. For example, how many vehicle kilometres are expended each weekend in the pursuit of children's sport? Could we quantify the environmental costs of divorce in terms of the expanded number of households? (Yu and Liu 2007). How many Australian children have two bedrooms as a consequence of separated parents?

How households entertain themselves similarly remains ignored in research on household sustainability. Entertainment should be considered an essential element of human existence: we need food, shelter, clothing *and* also pleasure. Music, art, dance, acting, having sex, eating for taste not just nutrition, socialising, gossiping

---

3   V. Organo, L. Head and G. Waitt, 'Who does the work in sustainable households? A time and gender analysis', in preparation.

and laughing are part of all functioning human societies across times and places. Because how we gain pleasure – how we entertain ourselves – is a basic part of life, it thus ought to be a crucial part of the climate change equation.

In 'Tough Times? Green Times?' we asked participants how they entertained themselves, how often they went out, what they did, and whether this had changed (and why). Nearly half the respondents had recorded a noticeable change in going out for entertainment in the last 12 months – of these three-quarters had reduced their social outings. Of the reasons given for changes in social outings, climate change was the least common (about 2 per cent), compared with the global financial crisis (25 per cent), having recently retired and health concerns (about 20 per cent each), unemployment, moving closer to or further away from friends, and having a baby (each about 10 per cent). As a consequence, households stayed home more often, ate out less and spent less money on the cinema, festivals, live theatre or music. This so-called cocooning has potential implications for sustainability given that that research which has sought to account for the ecological footprint of mass entertainment – principally festivals (Gibson and Wong 2011) – has shown that there are contradictions that play out in the specifics of cultural practices. Staying at home more means fewer trips in cars, reducing fuel dependency, but also atomizing modes of entertainment whereby separate household units need energy to watch television, and to light and heat their homes. At festivals (and presumably too at cinemas, theatres and live music venues) energy use is by contrast borne collectively, meaning lower per capita carbon emissions.

These dilemmas of social practice become even more complicated if we then analyse the meaning of eating out socially, the contributions festivals make to public culture, or the thrill of the eBay™ hunt and its associated sociality. They take us to the heart of how symbols, habit, knowledge and practice are entwined in our daily lives.

Conclusion

In this chapter we have sought to respond to Mike Hulme's (2008) challenge to enculturate climate change research, with particular regard to governmental compulsions towards 'green' behaviour amongst citizens. As the hub of domestic consumption, households are increasingly encouraged to consider and reduce their greenhouse gas emissions. But as we have sought to show here, the household as a site of material cultural relations is far more complex than policy sound-bites about the 'green home' might suggest. The burdens and the productive possibilities associated with such reductions are unevenly distributed spatially into future times and places. Our survey and ethnographic work in Wollongong challenges long-held ideas that cities, households or individuals categorised as working class are less likely to be engaged in 'environmental' action. Changing patterns of household behaviour, across a socio-economic continuum, were driven more often by cost and changing household social circumstances, rather than concerns about climate change. But climate change is part of the mix.

This work also unsettles the idea that consumption itself is 'bad'. For those who are poor, consumption equals survival, while the desire to move beyond imminent crisis into consumption for luxury remains a strong – if likely unfulfilled – emotional pull. How willingly will people accept 'being greener and poorer', given that sustainability assumes a certain amount of 'loss' of lifestyle and material ownership (Humphery 2010)? And if they reject it, how might this translate into politics, meaning perhaps that governments take retrograde steps, or at best run platforms of tokenistic, techno-centric solutions rather than real change? Rather, we need to think through more nuanced ways in which consumption in interaction with other social and industrial processes (Shove 2003) produces dilemmas that are difficult to resolve; we may need to accept that there are multiple readings of sustainability for households.

References

Askew, L. and McGuirk, P. 2004. Watering the suburbs: distinction, conformity and the suburban garden. *Australian Geographer*, 35(1), 17–27.

Barr, S. 2003. Strategies for sustainability: citizens and responsible environmental behaviour. *Area*, 35(3), 227–240.

Bassett, T.J. 2010. Slim pickings: fair-trade cotton in West Africa, *Geoforum*, doi:10.1016/j.geoforum.2009.03.002 , 41 (1), 44–55.

Beard, N.D. 2008. Branding of ethical fashion and the consumer: a luxury niche or mass-market reality? *Fashion Theory: The Journal of Dress, Body and Culture*, 12(4), 447–468.

Blunt, A. 2005. Cultural geographies of home. *Progress in Human Geography*, 29(4), 505–515.

Blunt, A. and Dowling, R. 2006. *Home.* London and New York: Routledge.

Borger, E (2010) Dynamics of Extended Family Households: A Cultural Economy of Sustainability, BSc Honours Thesis, University of Wollongong.

Burgess, J., Bedford, T., Hobson, K., Davies, G. and Harrison, C. 2003. (Un)sustainable consumption, in *Negotiating Environmental Change: New Perspectives from Social Science*, edited by I. Scoones and M. Leach, Cheltenham: Edward Elgar, 261–291.

Chatterton, P. 2010. The urban impossible: a eulogy for the unfinished city. *City*, 14(3), 234–244.

Choice Australia 2009. Test HD set-top boxes – HD: highly disappointing. *Choice*, February, 59.

Davison, A. 2008. Contesting sustainability in theory-practice: in praise of ambivalence. *Continuum*, 22(2), 191–199.

Doeringer, P. and Crean, S. 2006. Can fast fashion save the US apparel industry? *Socio–Economic Review*, 4(3), 353–377.

Dowling, R. 2000. Cultures of mothering and car use in suburban Sydney: a preliminary investigation. *Geoforum*, 31(3), 345–353.

Fletcher, K. 2008. *Sustainable Fashion and Textiles: Design Journeys*. London: Earthscan.

Gibson, C. and Stanes, E. 2010. Is green the new black? Exploring ethical fashion consumption, in *Ethical Consumption: A Critical Introduction*, edited by T. Lewis and E. Potter. London: Routledge, 169–185.

Gibson, C. and Wong, C. 2010. Greening rural festivals: ecology, sustainability and human-nature relations, in *Festival Places*, edited by C. Gibson and J. Connell. Bristol: Channel View, 92–105.

Gibson, C., Head, L., Gill, N. and Waitt, G. 2011. Climate change and household dynamics: beyond consumption, unbounding sustainability. *Transactions of the Institute of British Geographers*, no. doi: 10.1111/j.1475–5661.2010.00403.x, 36 (1), 3–8.

Gorman-Murray, A. 2006. Gay and lesbian couples at home: identity work in domestic space. *Home Cultures*, 3(2), 145–168.

Gough, J. 2010. Workers' strategies to secure jobs, their uses of scale, and competing economic moralities: rethinking the 'geography of justice'. *Political Geography*, 29(3), 130–139.

Gregson, N. and Beale, V. 2004. Wardrobe matter: the sorting, displacement and circulation of women's clothing. *Geoforum*, 35(6), 689–700.

Hamilton, C., Dennis, R. and Baker, D. 2005. *Wasteful Consumption in Australia*. Canberra: The Australia Institute.

Hawkins, G. 2006. *The Ethics of Waste: How We Relate to Rubbish*. Sydney: University of New South Wales Press.

Hinchliffe, S. 1996. Helping the earth begins at home: the social construction of socio-environmental responsibilities. *Global Environmental Change*, 69(1), 53–62.

Hobson, K. 2003. Thinking Habits into Action: the role of knowledge and process in questioning household consumption practices. *Local Environment*, 8(1), 95–112.

Hobson, K. 2006. Bins, bulbs, and shower timers: on the 'techno-ethics' of sustainable living. *Ethics, Place and Environment*, 9(3), 317–336.

Hulme, M. 2008. Geographical work at the boundaries of climate change. *Transactions of the Institute of British Geographers*, 33(1), 5–11.

Humphery, K. 2010. *Excess: Anti-Consumerism in the West*. Cambridge: Polity Press.

Johnston, J., Biro, A. and MacKendrick, N. 2009. Lost in the supermarket: the corporate-organic foodscape and the struggle for food democracy. *Antipode*, 41(3), 509–532.

Lane, R., Horne, R. and Bicknell, J. 2009. Routes of reuse of second-hand goods in Melbourne households. *Australian Geographer*, 40(2), 151–168.

Lash, S. and Urry, J. 1994. *Economies of Signs and Space*. London: Sage.

Lenzen, M., Dey, C and Foran, B. 2004. Energy requirements of Sydney households. *Ecological Economics*, 49(3), 375–399.

Malpass, A., Barnett, C., Clarke, N. and Cloke, P. 2007. Problematizing Choice: responsible consumers and sceptical citizens, in *Governance, Consumers and Citizens: Agency and Resistance in Contemporary Politics*, edited by M. Bevin and F. Trentmann. Basingstoke: Palgrave MacMillan, 231–246.

Power, E. 2009. Domestic temporalities: nature times in the house-as-home. *Geoforum*, 40(6), 1024–1032.

Reid, L., Sutton, P. and Hunter, C. 2009. Theorizing the meso level: the household as a crucible of pro-environmental behaviour, *Progress in Human Geography*, 34(3), 309–327.

Robbins, P. 2007. *Lawn People: How Grasses, Weeds, and Chemicals Make Us Who We Are.* Philadelphia: Temple University Press.

Saunders, C., Barber, A. and Taylor, G. 2006. *Food Miles – Comparative Energy/ Emissions Performance of New Zealand's Agriculture Industry*, AERU Research Report No. 285.

Shove, E. 2003. *Comfort, Cleanliness and Convenience.* Oxford: Berg.

Shove, E., Watson, M., Hand, M. and Ingram, J. 2007. *The Design of Everyday Life.* Berg: Oxford.

Yu, E. and Liu, J. (2007). Environmental impacts of divorce. Proceedings of the National Academy of Sciences 104, pp. 20629–20634.

Chapter 3

A Domestic Twist on the Eco-efficiency Turn: Environmentalism, Technology, Home

Aidan Davison

Introduction

Environmentalist engagements with questions of technology have changed and diversified a good deal in high consumption societies over the past three decades. These changes bear witness to the institutionalising of environmental concerns and the proliferation of competing environmentalist identities (Grendstad et al. 2006, Pakulski and Tranter 2004, Dryzek et al. 2003). At the centre of these changes is the 'eco-efficiency turn' that took place in environmental politics during the 1990s, enabling a new alignment of ideals of progress and sustainability. This alignment underpins the reinvention of capitalism as a leading force for sustainable production (Davison 2001, 2008).

In this chapter, I provide an account of the eco-efficiency turn in English-speaking high consumption societies, paying particular attention to domestic life. A focus on domestic life may seem incongruous, for the goal of eco-efficiency has been driven by corporations and centred on matters of production. Yet market-led pursuit of sustainable production has advanced in lock-step with neo-liberal agendas for sustainable consumption that emphasise individual responsibility. Central to these agendas are public policies and environmentalist tactics that foreground the role of householders in taking up more eco-efficient goods and services. In this chapter, I place emerging interest in the eco-efficient home in the context of earlier environmentalist engagements with the technology of home and consider how the goal of eco-efficiency may be transformed by its entry into the domestic sphere.

The chapter is in two parts. The first sketches out the eco-efficiency turn. Recent agendas for sustainable production and sustainable consumption are set against the alternative technology movement, whose heyday was the 1970s. This movement's goal of domestic self-sufficiency appears very different from that of eco-efficiency. However both goals are founded on similar and problematical assumptions about the role of technology in political change. I aim to take seriously Langdon Winner's (1986: 39) observation that "in our times people are often willing to make drastic changes in the way they live to accommodate technological innovation while at the same time resisting similar kinds of changes justified on political grounds." This apparent disparity between the technological malleability and political intractability of domestic practices draws attention to the ways in which technology operates as implicit, de facto politics in contemporary societies. Indeed, a premise of this chapter is that analysis of technology as de facto politics sheds light on current efforts to materialise the goal of sustainability.

Drawing on this analysis, the second part of the chapter focuses on the entry of agendas for eco-efficiency into Australian homes – spaces that truly are, as the Your Home program, a partnership of government and industry, observes, 'much more than a place to live' (www.yourhome.gov.au/renovatorsguide/getting-started.html). Tracing key features of the culture of home in Australian society through themes of retreat and renovation, I ask how the goal of the eco-efficient home promises to confound the instrumentalist assumptions about the neutrality of technology that underpin the eco-efficiency agenda. Drawing on Fiona Allon's (2008) account of the interplay of domestic practices and Australian nationhood, I conclude that efforts to encourage householders to take responsibility for sustainability through domestic technologies will give rise to paradoxical results.

I. From Self-sufficiency to Eco-efficiency

The United Nations 1992 Earth Summit marks the beginning of the eco-efficiency turn (Dryzek 1997). In the three decades before this, the 'environment' was caught up in a counter-cultural critique of modernity in high consumption societies. Defending a fragile planet against insatiable technology, many environmentalists sought personal refuge from the perceived pathology of modern societies in the purity of wilderness, the authenticity of tradition and the new age of post-secular spirituality. Fashioned out of the postwar boom, counter-cultural environmentalism grew in influence through the 1970s and early 1980s. Combined with the rising science of ecology, environmental movements drove concerns about nature into legal, economic, educational and political institutions. Fritjof Capra's optimism in *The Turning Point* (1982) that modern society was collapsing under the weight of hubris, making way for a newborn, holistic civilisation, was shared widely in environmental movements. These movements combined talk of epochal shifts in worldviews and social paradigms with an emphasis on personal virtues of simplicity and self-realisation. Nature was taken to be an objective arbiter of social affairs and a subjective participant in personal transformation. Reflecting this, the new field of environmental studies encompassed the managerial professions of nature conservation and radical philosophies of ecological selfhood.

It was widely assumed in counter-cultural environmentalism that nature's commandments provided a self-evident blueprint for the political transformation of modernity. Confident that age-old struggles for power would be trumped by the science of ecology, many turned their attention to immediate matters of everyday life, finding in nature instructions for living (Peterson del Mar 2006). Self-sufficiency, voluntary simplicity, vegetarianism, bioregionalism and organic design were prominent among the themes that aligned environmental politics with a personal ethos of alternative living (for example Sale 1985). The appropriate or alternative technology movement – the former term being prevalent in North America, the latter in the United Kingdom, with both appearing in Australia – was at the centre of efforts to translate environmentalism into a way of life. This movement put its faith in a new kind of technology; one characterised by smallness, softness, simplicity, localness and conviviality (Smith 2005, Kleiman 2004, Pursell 1993, Willoughby 1990, Winner 1986).

As evident in its manifesto, Fritz Schumacher's *Small is Beautiful*, the alternative technology movement formed around issues of 'third world' development in the 1960s. During the 1970s, however, Schumacher's (1973: 128) vision of technology that 'is conducive to decentralisation, compatible with the laws of ecology, gentle in its use of scarce resources, and designed to serve the human person instead of making him the servant of machines', was taken up by environmentalists in high consumption societies as an agenda for self-transformation (Willoughby 1990). Emphasis on decentralisation and domestic autonomy fitted with environmentalist arguments that modern institutions were on the brink of collapse, arguments that gained force from the oil shock of 1973 (for example Vale and Vale 1975). Rather than inciting protest against powerful institutions, the alternative technology movement advocated freedom from the self-destructive chaos of modernity in the form of an intrinsically satisfying pursuit of domestic self-sufficiency. As Winner (1986: 76) puts it, the logic of the movement went thus: 'Who needs doctors? Do your own health care. Who needs architects and builders? Build your own home. Who needs utilities? Generate your own energy'. This celebration of individual autonomy was not viewed as an expression of the liberal individualism of which environmentalism was so critical. Rather, according to Winner (1986: 79), the technology choices of individuals were seen to form the straightest path to a truly post-materialistic political order:

> A person would build a solar house or put up a windmill, not only because he or she found it personally agreeable, but because the thing was to serve as a beacon to the world, a demonstration model to inspire emulation. If enough folks built for renewable energy, so it was assumed, there would be no need for the nation to construct a system of nuclear power plants. People would, in effect, vote on the shape of the future through their consumer/builder choices … Radical social change would catch on like disposable diapers, Cuisinarts, or some other popular consumer item.

Yet the self-sufficiency advocated by alternative technologists did not catch on in high consumption societies. By the mid-1980s, this movement had all but collapsed in North America (Pursell 1993) and was in decline in the United Kingdom (Smith 2005), although it continues to be influential in social movements focused on low consumption societies.

The decline of the alternative technology movement's influence in high consumption societies is linked, ironically, to the success of environmentalism in having modern institutions take up scientific concerns about the environment during the 1980s. This success saw environmental issues reinvented as an instrumental challenge; a stimulus for technological innovation and economic growth (Davison 2001). Stripping away counter-cultural projects, modern institutions were left with technical matters of resource use and environmental services in relation to which talk of paradigm shifts seemed irrelevant and an emphasis on personal lifestyle inadequate. Having predicted this eventuality in the 1960s, the eco-anarchist Murray Bookchin (1982: 14) penned this warning to the North American environmental movement in the early 1980s:

> The hoopla about a new 'Earth Day' or future 'Sun Days' or 'Wind Days', like the pious rhetoric of fast-talking solar contractors and patent hungry 'ecological' inventors, conceal the all-important fact that solar energy, wind power, organic agriculture, holistic health, and 'voluntary simplicity' will alter very little in our grotesque imbalance with nature if they leave the patriarchal family, the multinational corporation, the property system, and the prevailing technocratic rationality untouched.

The appropriate technology movement shied away from explicit political argument, combining the apolitical authority of nature with the technologist's fascination for design. Alternative technologies were offered up as being more efficient for being more natural than those of modernity, working with rather than against natural laws of design. The political limitations of this strategy are evident in the collapse of this movement as it became apparent that cherished tools and techniques – renewable energy and organic agriculture, in particular – were as suited to centralised control and capital accumulation as they were to bioregional decentralisation and steady-state economics. Writing in 1983, Winner (1986: 54) anticipated this collapse with characteristic lucidity:

> Because the idea of efficiency attracts a wide consensus, it is sometimes used as a conceptual Trojan horse by those who have more challenging political agendas they hope to smuggle in. But victories won in this way are in other respects great losses. For they affirm in our words and in our methodologies that there are certain human ends that no longer dare be spoken in public. Lingering in that stuffy Trojan horse too long, even soldiers of virtue eventually suffocate.

The rapid suffocation of counter-cultural hope for a technology-led paradigm shift was matched by an equally rapid rise in the confidence of modern institutions that they could direct technological means to environmental ends. This enthusiasm was at the core of the concept of sustainable development promoted so effectively by the 1987 'Brundtland' *Report of the World Commission on Environment and Development*. Built on the two-step claim that economic growth is essential to overcoming poverty and that ecological life-support systems are essential to sustaining economic growth, this concept was quickly dismissed by counter-cultural environmentalists as an oxymoron and just as quickly adopted by governments (Davison 2001). Displacing emphasis on the absolute limits of nature, the Commission wrote not of 'absolute limits but limitations imposed by the present state of technology and social organization on environmental resources' (WCED 1987: 8).

The central task of the 1992 Earth Summit was to translate sustainable development into an agenda for international action. However, during two years of preparatory negotiations, tensions between wealthy and poor nations ran high. While wealthy nations placed population growth at the centre of debate about sustainable development, poor nations pointed to inequality in resource consumption. The Summit's resulting action plan, *Agenda 21*, failed to resolve this tension, binding conflicting goals with motherhood aspirations. Although influential at the level of municipal government, *Agenda 21* produced little meaningful reform on the part of the national governments (Davison 2001). This is not to say that the Earth Summit achieved little. Reflecting the impact of neo-liberal reforms, it was not nation-states but businesses and consumers who emerged from the Summit as leading the pursuit of sustainable development.

Business came of age as an advocate of sustainable development at the Earth Summit, most notably in the form of the (World) Business Council for Sustainable Development, a group described by the Summit's Secretary-General as 'a cadre of the world's leading practitioners of sustainable development' (Strong, cited in Chatterjee and Finger 1994: 117). Comprised of the leaders of many of the world's largest transnational corporations, this group was asked in the lead-up to the Summit to 'develop its contribution, both in word and deed, to sustainable development' (WBCSD 2000: 9). Swiss industrialist Stephan Schmidheiny explains that, in word, this contribution took the form of a new concept.

> In 1991, we in the then Business Council for Sustainable Development were looking for a single concept, perhaps a single word, to sum up the business end of sustainable development. Finding no concept on the lexicographer's shelf, we decided we would have to launch an expression. After a contest and much agonizing, we came up with eco-efficiency. In simplest terms, it means creating more goods and services with ever less use of resources, waste and pollution. (Schmidheiny in WBCSD 2000: Forward)

Using this concept, the Business Council argued for business self-regulation, teaming up with the International Chamber of Commerce during the Summit preparations to thwart efforts by the UN Centre for Transnational Companies to develop a regulatory code for transnational corporations. These efforts were so successful that the Centre was closed down; a fact some have linked to the Council's contribution to the Summit in deed, which was to provide significant sponsorship to a cash-strapped United Nations (Chatterjee and Finger 1994: 28–30, 114–117).

The concept of eco-efficiency offered to bypass adversarial debate about population growth and consumption. By the mid-1990s, optimism had grown amongst governments and in some environmental movements that a 'factor 4' increase in eco-efficiency was feasible, enabling, in theory, a population twice the size of today's to enjoy a high standard of living while halving existing resource demand and waste production (von Wiezsäcker, A.B. Lovins and L.H. Lovins 1997). By the end of the 1990s, the influential book *Natural Capitalism* asserted that business could lead a 'factor 10' improvement in eco-efficiency. In contrast to the earlier conviction of the alternative technology movement that centralised technological systems were headed for collapse, *Natural Capitalism* (Hawken, A.B. Lovins and L.H. Lovins 1999: 1–2) placed utopian faith in the power of capitalism to harness technology to the goal of sustainable development:

> imagine for a moment a world where cities have become peaceful and serene because cars and buses are whisper quiet, vehicles exhaust only water vapour …
> Living standards for all people have dramatically improved, particularly for the poor and those in developing countries. Involuntary unemployment no longer exists, and income taxes have largely been eliminated … The frayed social nets of Western countries have been repaired …
> Is this the vision of a utopia? In fact, the changes described here could come about in the decades to come as the result of economic and technological trends already in place.

One of the authors of *Natural Capitalism*, and of *Factor Four*, Amory Lovins (1977: 150–151), had been a leading figure in the alternative technology movement during the 1970s. At this time, Lovins advocated 'soft energy systems [that] have an obvious relevance to everyday life because they are both physically and conceptually closer to end users', in contrast to 'hard technologies … oriented toward abstract economic services for remote and anonymous consumers'. As implied in this quotation, the alternative technology movement was generally dismissive of consumers. In the spirit of Robert Pirsig's 1974 *Zen and the Art of Motorcycle Maintenance*, alternative technologists valued a life of self-sufficiency characterised by problem-solving, manual skill and rugged independence over a passive life of consumption characterised by banal talents, flimsy pleasures and dependence on experts. Lovins' latter-day faith in capitalism, then, is indicative of a wider rehabilitation of consumption practices in environmental politics.

Like 'eco-efficiency', the concepts of sustainable consumption and sustainable production debuted on the world stage at the Earth Summit. The initial function of these twin terms was to highlight the environmental dangers of overconsumption. Reflecting the concern of low consumption nations, *Agenda 21* declared 'the major cause of the continued deterioration of the global environment is the unsustainable pattern of consumption and production, particularly in industrialized countries' (UN 1992: 4.3). By the mid-1990s, the United Nations was referring to 'sustainable consumption as an "overriding issue" and a "cross-cutting theme" in the sustainable development debate' (Jackson 2006: 3). However by this time emphasis was shifting from overconsumption – and counter-arguments about overpopulation – to the eco-efficiency agenda. Talk of sustainable production and consumption referred less to reduced consumption than it did to the creation of environmental goods and services. The concept of sustainable production now reflected corporate confidence that voluntary sustainability initiatives were good for business. The concept of sustainable consumption reflected consumer confidence that, as Michael Maniates (2002: 45) has put it, 'knotty issues of consumption, consumerism, power and responsibility can be resolved neatly and cleanly through enlightened, uncoordinated consumer choice'.

Hope that eco-efficient consumption promises social transformation without political upheaval is reminiscent of the short-lived dreams of the alternative technology movement. Advocates of sustainable consumption and alternative technology both focus on the voluntary technology choices of householders. Where alternative technologists did so in the belief that modernity was collapsing around them, advocates of sustainable consumption do so with the support of neo-liberal reforms that have eroded government functions and increased the reach of market mechanisms. Both agendas have assumed that political objectives can be smuggled into the home in the guise of technical specifications and consumer preferences. In the second part of this chapter I focus on domestic practices in more detail by considering how the complex politics of home in Australia may give efficiency-led agendas for sustainable consumption a decidedly domestic twist.

II. Sustainability at Home in Australia

The 'Great Australian Dream' of home ownership sits at the centre of modern Australian nationhood. Frederic Eggleston (1932: 330–331), Attorney-General of Victoria during the 1920s, was right to observe that Australian democracy is founded upon 'self-contained man': a landholder whose 'hedges are his frontiers'. At Federation, in 1901, half of all Australian dwellings were owner-occupied (Butlin, cited in Davison 2006). By 1961, this proportion had risen to 70 per cent: at that point the highest level of home ownership in the world (Troy 2000: 719). Coupled with this achievement is Australia's status as one of the world's most suburbanised nations; even, arguably, the world's first truly suburban nation (Davison 2006). One of the distinguishing features of Australian suburbanisation is that it was, from the beginning, as much a proletarian as a bourgeois phenomenon. Indeed, until the middle of the twentieth century, the self-contained lives of Australia's large 'suburban peasantry' (Mullins 1981) bore more than passing resemblance to the vision of self-sufficiency that later inspired the alternative technology movement. Before 1950 the suburban home, a primary locus of women's labour, was often a site of resource autonomy, with water collection, food production and processing, on-site waste management, furniture and clothes making, and building construction and maintenance being commonplace (Troy 2003, Mullins 1981).

Australian home ownership levels have declined from their peak in the mid-1960s, on the back of neo-liberal reforms, and the home is today more likely given over to market-based consumption than to domestic production. However as Fiona Allon demonstrates in her recent book, *Renovation Nation*, the political influence of the dream of a nation founded upon autonomous householders remains strong. Tracing the bonds that join the home to the nation, Allon (2008: 2) observes that throughout modern Australian history both home and nation have been 'fortresses inside which we worried about safety and security and protecting our wealth'. Domestic practices have constituted a material arena in which immediate, tangible concerns have been translated into the symbolic work of nation-building. Built on British invasion and at a continental scale, the abstractions of the Australian nation-state have been grounded in the sovereign home. The reality of home and garden – the 'permanence and authenticity' of an undeniably real estate (Allon 2008: 62) – has provided ontological as well as material security for many. Writing between stints as Prime Minster, in the midst of the Second World War, Australia's most influential champion of home ownership, Robert Menzies (1992 [1942]: 7–8) lauded the suburban desire 'to have one little piece of earth with a house and a garden which is ours, to which we can withdraw, in which we can be among friends, into which no stranger may come against our will'. Honed in everyday life, this desire for domestic autonomy continues, ironically, to play a key role in Australian solidarity.

While the 'home-centredness' of Australian society is not new, Allon (2008: 36) claims that many Australians are presently embarked upon a new and intense period of '"cocooning" – a withdrawal into the security of the home' characterised by home renovation, and linked, among other things, to fears about 'illegal' immigration, border security and terrorism. As evidence of this cocooning she points outs that that nine out of ten Australian homeowners are engaged in improving their homes, with renovation accounting for A\$28.1 billion of economic activity in 2006–2007 (Allon 2008: 26, 39). Allon presents the willingness of Australians to repeatedly reinvent their homes as being characteristic of a society perpetually inventing itself in an attempt to find itself. Although allied to a long-standing Australian tradition of home-building and 'do-it-yourself' home maintenance by homeowners (Frost and Dingle 1995), the present preoccupation with renovation reveals a good deal about how the exercise of domestic autonomy has shifted from productive labour to self-expression through consumer choice. Nonetheless, like earlier phases of Australia's householder democracy, the culture of renovation responds to challenges faced by the nation through the exercise of domestic autonomy. 'There is,' after all, 'nothing like renovations to keep the world at bay: they focus our concentration on the here and now, the little details, and they appear to bring everything under our control' (Allon 2008: 13).

Allon argues that retreat into the security of home only heightens fear of an unruly world by cultivating a private experience of isolation and helplessness in the face of collective challenges. This in turn strengthens the domestic impulse to withdraw from external threats. Take, for example, the recent surge in private mortgage debt in Australia, from 20 per cent of GDP in 1991, to 85 per cent of GDP in 2008 (Keen 2009). Bearing out the Australian fascination with home ownership as a form of wealth creation, superannuation and bequest, this growth in private debt is linked, among other things, to taxation policies that encourage speculative private investment in housing and to the recent renovation boom. This growth in private debt has also seen Australian electoral politics more strongly tethered to short-term concerns about household budgets, especially in relation to mortgage interest rates. In order to purchase and perfect their homes, many Australians appear confident in committing an ever larger proportion of their incomes to servicing debt for decades into the future. This private confidence in the future stands in contrast, however, to the increasing myopia of electoral politics and the related timidity of many public policy responses to intergenerational issues. In particular, potentially grave risks posed by this surge in private debt to the future security of the Australian economy gain little political traction (Keen 2009). Indeed, talk of such risks may only further focus the attention of householders on the task of shoring up the home.

I have elsewhere argued that the paradoxical, self-defeating character of private retreat into a suburban idyll lies at the centre of postwar environmental politics in Australia (Davison 2006, 2008, 2009). The suburbs have been roundly criticised by environmentalists as the prime embodiment of unsustainable consumption in Australia. This critique has failed to provoke sufficient self-reflection on the part of many environmentalists about their own, often suburban, everyday lives. This critique has also largely failed to account for the origins of the Australian suburban home in what Zygmunt Bauman (1991) calls the ambivalence of modernity. The creation of Australian modernity has relied on the ability of a majority to shuttle back and forth between a (masculinised) modern world ruled by dispassionate reason and efficient technology and a (feminised) domestic idyll characterised by sentiment, tradition, religion and nature. Alexander Sutherland's 1890 (143) poem, *Home and the World*, is one in a long lineage of Australian tributes to the sanctuary of home:

> On the one side swelling from the petty strife
> Of men and business care;
> But on the other, where
> Thy Home extends the smoothness of its breast
> Sinking in trustful rest

Many Australian environmentalists appear to assume that domestic market-led consumption – for example related to gardening, food consumption, animal keeping, outdoor recreation and the intimate life of the body – are a straightforward expression of the modern will to dominate nature (Davison 2008). In my view, however, the emotive and embodied engagements with nature bound up in domestic consumption practices are better understood as existing in dialectical tension with the progress of modern reason: they exist, that is, as a necessary counterpoint to dispassionate techno-economic institutions that seek to control an instrumental nature. Reflecting this dialectic, generations of Australians have retreated to their own 'private Eden' at the limits of the city (Davison 2006), in the process pushing these limits further out, creating vast cities, underwriting techno-economic growth, and trapping earlier suburban pioneers behind expanding brick and tile frontiers. The failure of suburban lives to match the domestic ideal that motivated them has only further heightened a desire for this ideal, giving rise to many apparently anti-suburban expressions of suburban desire (Gilbert 1988). Consider, for instance, the present pursuit of domestic 'seachange' and 'treechange' in the ever-growing commuter belts and e-worker hinterlands of Australian cities, or the pursuit of vertical fringes of Australian inner-cities in spacious gated apartments.

In addition to fears about border security and terrorism identified by Allon, environmental fears are significant in the self-reinforcing domestic introversion she describes. After being in steady decline from 1990 to the early 2000s, public concern about the environment is again on the rise, driven, in particular, by drought and, more recently, by public acceptance of climate change and its implications for freshwater resources, bushfire risk and biodiversity conservation (Tranter 2010). These issues, and related concerns about food, water and energy security, have also been prominent in Australian political debate over recent years. After the failure of the legislated 1992 National Strategy for Ecologically Sustainable Development to bring about meaningful governmental reform – a failure linked to the rise of neo-liberal agendas for sustainable consumption described earlier – current policy responses to environmental issues in Australia have placed strong emphasis on the role of self-regulating consumers in achieving sustainable development. This emphasis builds on the long-established role of the self-contained householder in Australian politics. One result of this emphasis has been reliance on incentive schemes and education programs over regulatory strategies in governmental attempts to reduce household consumption of water, energy and other resources (for example Fielding et al. 2009, Productivity Commission 2005).

Often delivered through the business and non-governmental sectors, current governmental attempts to promote home sustainability in Australia are targeted at the uptake of eco-efficient technologies (see for example www.LivingGreener. gov.au). Rather than focusing directly on behavioural change, rebates, labelling schemes, low-interest financing and other public policy strategies seek to influence consumer choices in the knowledge that a great many Australians are actively engaged in construction and ongoing renovation of their homes. Considerations of eco-efficiency seem likely to become an increasingly routine aspect of householders' deliberations about how to perfect their homes. In addition to water and energy efficiency rating schemes for conventional domestic appliances and for buildings as a whole, there is presently a government-sanctioned uptake of 'green' domestic technologies, such as rainwater tanks, grey water recycling systems and photovoltaic arrays. Only a few years ago these and other eco-technologies marked out the homes of radical environmentalists in Australian cities. Today, techniques and tools for harvesting, recycling and conserving resources in the home – everything from solar passive architecture to sub-soil irrigation to paint with insulating properties – are increasingly bound up with conventional advertising, marketing and retailing practices. The achievement of green distinction in the home and garden is becoming a depoliticised consumer endeavour related to prevalent social norms about 'the good home'. Potentially overwhelming problems about climate change, the end of cheap fossil fuels and the availability of freshwater in the Earth's driest continent are being translated into a familiar Australian preoccupation with protecting the home from a dangerous world. Public anxieties about environmental unsustainability contrast sharply with the empowerment experienced by homeowners as they 'do their bit' for the environment (see Chapter 11 in this volume) through domestic consumption practices.

By way of concluding this chapter, I briefly focus on a recent Australian home sustainability initiative that sheds light on the changing environmental politics of domestic technology. This example of the convergence of domestic introversion and eco-efficient consumption centres on the Alternative Technology Association's (ATA) magazine *Sanctuary – Sustainable Living with Style* (see www.sanctuarymagazine.org.au). Formed in 1980, ATA is one of few remaining legacies of the Australian appropriate technology movement: the Centre for Appropriate Technology (CAT), formed in the same year, being another. Whereas CAT was and remains focused on remote Indigenous communities, ATA can rightly claim to be the 'leading not-for-profit organisation, promoting sustainable technology and practice' to Australian householders (see www.ata.org.au). The Australian alternative technology movement has not been documented in the social science literature. In the absence of such data, my involvement in Australian environmental movements and environmental studies suggests that the decline in the alternative technology movement observed in the northern hemisphere holds true for Australia, although both the rise and fall of this movement occurred later in Australia.

The continued existence of ATA gives the appearance that the appropriate technology movement persists. On closer inspection, however, this organisation reveals something of the way in which the appropriate technology movement was disabled by the splitting apart of technological and political aspirations. As noted above, the technologies long advocated by ATA, such as small-scale renewable energy, are no longer 'a fringe occupation', being of interest to 'millions of Australians' (ATA 2010: 40). Recent consumer interest in home sustainability has seen ATA gain wider relevance and support. At the same time, however, the counter-cultural vision on which ATA was founded – a vision of political decentralisation, steady-state economics, bioregional self-sufficiency and post-materialistic values – has largely given way to neo-liberal agendas for sustainable consumption.

The inaugural 1980 issue of ATA's first publication, *Soft Technology – Alternative Technology in Australia*, declared the organisation's aim to be the provision of 'practical information on what people have actually done in the area of alternative technology, and [to] show with the aid of diagrams how other people can build the same equipment themselves' (ATA 2007a: 52). The fading of a stark counter-cultural distinction between small, soft alternative technology and big, hard modern technology is reflected in the renaming of this magazine in 1997. Under the title *ReNew - Technology for a Sustainable Future* this magazine continues to appeal to ATA's core membership, a group of technically adept environmentalists, early adopters of 'cutting edge technologies and practices' (ATA 2009b: 3), although advertisements promoting expert knowledge and products are often more prominent than backyard builders. In contrast, ATA's second magazine, *Sanctuary*, launched in 2007, is squarely aimed at the emerging constituency of 'environmentally conscious consumers' (ATA 2009a: 7) who wish to 'future-proof' (ATA 2008: 42) their 'sustainable dream homes' (ATA 2010: 39).

Launched with the support of the Your Home program – a 'joint initiative of the Australian Government and the design and construction industries' (www. yourhome.gov.au) – *Sanctuary* is a 'stylish title' focused on environmental home renovation and building. Providing generous space to advertisers, this photograph-laden magazine targets 'educated urban professionals who care about the environment but do not want to compromise on style and comfort', and who 'are not constrained by budget' (ATA 2009a: 2–3). The founding premise of *Sanctuary*, as laid out in its first editorial, is that the sustainable home is 'no longer simply the domain of ... "alternative" lifestylers': indeed, readers are assured that 'a sustainable home will actually increase your level of comfort and ease the strain on your hip pocket' (ATA 2007b: 6). Bearing out ATA's (2009a: 2) claim that 'consumers are hungry for the information and products' related to home sustainability, subscriptions to *Sanctuary* increased 51 per cent during 2009, resulting in an estimated 70,000 readers. In addition to its title – and the curious invitation to 'sanctify your home' by finding out about eco-efficient technologies offered by the magazine's website – the feature stories on individual homes and householders that make up almost all of *Sanctuary*'s non-commercial content speak directly to Australia's history of domestic introversion. Take, for example, this excerpt from an architect's brief provided by an inner-suburban couple who paid a substantial sum to create an eco-efficient backyard studio:

> Just erect me a tin shed of modest comfort;
> Where I can feel a happiness without cause
> and shut out the oily businesses of the world. (ATA 2009c: 49)

My aim in this brief example has not been to disparage endeavours such as *Sanctuary* magazine, although this chapter has, more generally, contested the neo-liberal agenda of sustainable consumption to which it belongs. I have used this example to illustrate, first, the importance of Australia's tradition of domestic introversion in the emerging preoccupation with domestic eco-efficiency and, second, the continued failure of environmental movements to realise their political ends through strategies that regard technology merely as apolitical means. The technology of home is tangled up in Australia in the complex politics of nationhood and does not resemble the instrumental tools and techniques imagined in agendas of eco-efficiency. One householder featured in *Sanctuary* unwittingly gets to the heart of the matter in suggesting that 'our dwellings reflect our social conscience' (ATA 2007b: 9). It seems equally plausible to suggest that the social conscience of Australians is, in large part, to be found in the home.

References

Allon, F. 2008. *Renovation Nation: Our Obsession with Home*. Sydney: UNSW Press.

ATA, *see* Alternative Technology Association

Alternative Technology Association. 2007a. *ReNew*, 100. [Online]. Available at: http://shop.ata.org.au/cart.php?target=product&category_id=300&product_id=16464 [accessed 15 September 2010].

Alternative Technology Association. 2007b. *Sanctuary*, 1.

Alternative Technology Association. 2008. *Sanctuary*, 3.

Alternative Technology Association. 2009a. *Sanctuary: Media Kit 10*. [Online]. Available at: www.sanctuarymagazine.org.au/documents/Sanctuary_Media_Kit.pdf [accessed 10 May 2010].

Alternative Technology Association. 2009b. *Alternative Technology Association Strategic Plan 2009–2012*. [Online]. Available at: www.ata.org.au [accessed 20 March 2010].

Alternative Technology Association. 2009c. *Sanctuary*, 9.

Alternative Technology Association. 2010. *Sanctuary*, 10.

Bauman, Z. 1991. *Modernity and Ambivalence*. Ithaca, NY: Cornell University Press.

Bookchin, M. 1982. An open letter to the ecological movement. *Social Alternatives*, 2(3), 13–16.

Capra, F. 1982. *The Turning Point: Science, Society and the Rising Culture*. London: Flamingo.

Chatterjee, P. and Finger, M. 1994. *The Earth Brokers: Power, Politics and World Development*. London and New York: Routledge.

Davison, A. 2001. *Technology and the Contested Meanings of Sustainability*. Albany, NY: State University of New York Press.

Davison, A. 2006. Stuck in a cul-de-sac? Suburban history and urban sustainability in Australia. *Urban Policy and Research*, 24(2), 201–216.

Davison, A. 2008. The trouble with nature: ambivalence in the lives of urban Australian environmentalists. *Geoforum*, 39(3), 1284–1295.

Davison, A. 2009. Living between nature and technology: The suburban constitution of Australian environmentalism, in *Technonatures: Environments, Technologies, Spaces and Places in the Twentieth First Century*, edited by D. White and C. Wilbert. Waterloo, Canada: Wilfrid Laurier University Press, 171–193.

Dryzek, J. 1997. *The Politics of the Earth: Environmental Discourses*. Oxford: Oxford University Press.

Dryzek, J., Downes, D., Hunold, C., Schlosberg, D. with Hernes, H-K. 2003. *Green States and Social Movements: Environmentalism in the United States, United Kingdom, Germany, and Norway*. Oxford: Oxford University Press.

Eggleston, F.W. 1932. *State Socialism in Victoria*. London: P.S. King & Son.

Fielding, K.S., Louis, W.R., Warren, C. and Thompson, A. 2009. *Environmental Sustainability in Residential Housing: Understanding Attitudes and Behaviour towards Waste, Water, and Energy Consumption and Conservation among Australian Households*. Positioning Paper No. 12. Brisbane: Australian Housing and Urban Research Institute.

Frost, L. and Dingle, T. (1995) Sustaining suburbia: an historical perspective on Australia's urban growth, in *Australian Cities: Issues, Strategies and Policies for Urban Australia in the 1990s*, edited by P. Troy. Cambridge: Cambridge University Press, 20–38.

Gilbert, A. 1988. The roots of anti-suburbanism in Australia, in *Australian Cultural History*, edited by S.L. Goldberg and F.B. Smith. Melbourne: Cambridge University Press, 33–49.

Grendstad, G., Selle, P., Strømsnes and Bortne, Ø. 2006. *Unique Environmentalism: A Comparative Perspective*. New York: Springer.

Hawken, P., Lovins, A.B. and Lovins, H.L. 1999. *Natural Capitalism: The Next Industrial Revolution*. London: Earthscan.

Jackson, T. 2006. Readings in sustainable consumption, in *The Earthscan Reader in Sustainable Consumption*, edited by T. Jackson. London: Earthscan, 1–23.

Keen, S. 2009. Household debt – the final stage in an artificially extended Ponzi bubble. *Australian Economic Review*, 43(2), 347–357.

Kleiman, J. 2004. The Appropriate Technology Movement, in the *Encyclopedia of American Social Movements*, edited by I. Ness. Armonk, NY: M.E. Sharpe, 1317–1322.

Lovins, A. 1977. *Soft Energy Paths: Towards a Durable Peace*. Penguin: Harmondsworth.

Maniates, M. 2002. Individualisation: plant a tree, buy a bike, save the world?, in *Confronting Consumption*, edited by T. Princen, M. Maniates and K. Conca. Boston, MA: MIT Press, 43–66.

Menzies, R. 1992 [1942]. The forgotten people, reproduced in J. Brett, *Robert Menzies' Forgotten People*. Sydney: Pan Macmillan, 5–14.

Mullins, P. 1981. Theoretical perspectives on Australian urbanisation: I. Material components in the reproduction of Australian labour power. *Australian and New Zealand Journal of Sociology*, 17(1), 65–76.

Pakulski, J., and Tranter, B. 2004. Environmentalism and social differentiation. *Journal of Sociology*, 40(3), 221–235.

Peterson del Mar, D. 2006. *Environmentalism*. Harlow, UK: Pearson.

Pirsig, R. 1974. *Zen and the Art of Motorcycle Maintenance: In Inquiry into Values*. New York: William Morrow.

Productivity Commission. 2005. *The Private Cost Effectiveness of Improving Energy Efficiency*, Report no. 36, Canberra: Australian Government.

Pursell, C. 1993. The rise and fall of the appropriate technology movement in the United States, 1965–1985. *Technology and Culture*, 34(4), 629–637.

Sale, K. 1985. *Dwellers in the Land: The Bioregional Vision*. San Francisco: Sierra Club Books.

Schumacher, E.F. 1973. *Small is Beautiful: A Study of Economics as if People Mattered*. London: Abacus.

Smith, A. 2005. The alternative technology movement: an analysis of its framing and negotiation of technology development. *Research in Human Ecology*, 12(2), 106–119.

Sutherland, A. 1890. *Thirty Short Poems*. Melbourne: Melville, Mullen & Slade.

Tranter, B. 2010. Political divisions over climate change and environmental issues in Australia. *Environmental Politics* 20 (1), 78–96.

Troy, P. 2000. Suburbs of acquiescence, suburbs of protest. *Housing Studies*, 15(5), 717–738.

Troy, P. 2003. Saving our cities with suburbs, in *Dreams of Land: The Griffith Review*, edited by J. Schultz. Meadowbank, Qld/Sydney: Griffith University/ ABC Books, 115–127.

United Nations. 1992. United Nations Department of Economic and Social Affairs, *Agenda 21*. [Online]. Available at: www.un.org/esa/dsd/agenda21/ res_agenda21_04.shtml [accessed: 6 September 2009].

Vale, B. and Vale, R. 1975. *The Autonomous House: Design and Planning for Self-Sufficiency*. London: Thames and Hudson.

von Weizsäcker, E., Lovins, A.B. and Lovins, L.H. 1997. *Factor Four: Doubling Wealth-Halving Resource Use. A New Report for the Club of Rome*. Sydney: Allen & Unwin.

WBCSD, *see* World Business Council for Sustainable Development

World Business Council for Sustainable Development. 2000. *Eco-efficiency: Creating more Value with Less Impact*. [Online]. Available at: www.wbcsd. org/web/publications/eco_efficiency_creating_more_value.pdf [accessed: 3 September 2009].

WCED see World Commission on Environment and Development

World Commission on Environment and Development. 1987. *Our Common Future. The Report of the World Commission on Environment and Development*. Oxford: Oxford University Press.

Willoughby, K.W. 1990. *Technology Choice: A Critique of the Appropriate Technology Movement*. Boulder and London: Westview Press.

Winner, L. 1986. *The Whale and the Reactor: Limits in an Age of High Technology*. Los Angeles: University of California Press.

Chapter 4

Sustainability, Consumption and the Household in Developing World Contexts

Willem Paling and Tim Winter

To date, much of the debate, both public and academic, around household sustainability and the ethics of associated forms of consumption has primarily been situated within certain geographic and cultural contexts. To be specific, the idea of household sustainability, framed as it is and projected onto anxieties about climate change and global peril, has essentially been tied to the shifting moral landscape of consumption in the industrialised, or indeed post-industrial, societies of the so-called global North or 'West'. Far less attention has been given to household sustainability in the 'developing world'. Instead discussions around environmental impact have been oriented towards issues like urban pollution, the inadequacies of infrastructure, such as transport, waste management and so forth. Such points of focus remain prevalent because environmental concerns continue to be enmeshed in paradigms of 'development'; a discourse advanced during the second half of the twentieth century, defined by the goals of poverty alleviation, economic growth and rising GDPs. Understandably, the household has been seen as a key site of wealth accumulation vis-à-vis poverty reduction and as a marker of social mobility.

Householders in the 'developing world' are generally less aware of global issues. They tend to be less likely to be in a position where they can afford to consume at a level that has a significant impact on global climate change. Certainly the number of households that have been in a position to sustain high levels of consumption for multiple generations is low. In some respects it is only from a position of great privilege – often accompanied by high levels of consumption spread over multiple generations – that the luxury of concerning oneself with the effect of household practices on global climate change is practised. For Connolly and Prothero (2008), the 'green consumer' in the West can be understood through the lens of reflexive modernisation. The notion of reflexive modernity (see for example Beck, Giddens and Lash 1994, Beck 1992, Giddens 1991) holds that while the world remains modern – in spite of post-modern claims to the contrary – it is now reflectively looking back on itself in response to the actuality of modernity, parts of which were totally unanticipated by the ideologies of first modernity on which it was based. Viewed through such a lens, it is understandable that in the context of a relative lack of modernity in Cambodia – the focus of this chapter – such reflection and 'green consumerism' is not commonplace.

More recently, texts like *The New Consumers: The Influence of Affluence on the Environment*, by Norman Myers and Jennifer Kent, have been alerting us to the urgent need to address the rapidly rising levels of consumption in the developing world. The authors describe the key actors in this process as 'new consumers': 'people within an average of four-member households who possess purchasing power parity (PPP) of at least PPP $USD10,000 per year, or at least PPP $USD2500 per person' (2004: 8).

> By 2000 there were 1.1 billion of these people, with 945 million in 17 developing countries and 115 million in three transition countries. They totaled way more than the 850 million of the rich world, though with affluence generally far behind. They amounted to almost one-fifth of all citizens of the developing world, whereupon the label 'developing world' became a distinct misnomer. (Myers and Kent 2004: 15)

These 'new consumers' are typically:

> the long-standing members of the middle and upper class classes that can include senior managers, small business owners, investment bankers, physicians, lawyers, marketing executives, real estate agents, Internet engineers, architects, journalists, private school teachers, home designers, and insurance salespeople, [as well as] more recent members that can include computer programmers, junior managers, accountants, bank tellers, secretaries, and many others of similar status. (Myers and Kent 2004: 16)

Myers and Kent put forward a list of 'new consumer societies', which they populate with twenty countries. Among a range of analytic considerations, inclusion in this category requires a population exceeding 20 million, and an economic growth rate that exceeds five per cent per year.[1]

In examining the consumption practices of these new consumers they point to a number of distinct, upward trends, such as private car ownership, the daily consumption of meat, and the demand for electricity at the household level driven by a rapid growth in the number of energy intensive domestic appliances being purchased. Crucially, they argue that the number of 'the new consumers' will continue to 'soar' both in the medium and longer term. Asia, home to ten of the twenty new consumer economies, is rapidly becoming 'the center of gravity for the new consumer phenomenon' (Myers and Kent 2004: 18). Writing in 2004 they estimate the region will have around 900 million 'new consumers' by 2010; a figure

---

1   The twenty countries listed are as follows – in Asia: China, India, South Korea, Philippines, Indonesia, Malaysia, Thailand, Pakistan, Iran and Saudi Arabia; in Africa: South Africa; in Latin America: Brazil, Argentina, Venezuela, Colombia and Mexico; in Eastern Europe: Turkey, Poland, Ukraine and Russia (Myers and Kent 2004: 16).

that, despite the economic downturn of 2008–2009, seems plausible given the double digit growth in GDP for both China and India over much of this period.

As the title of their book suggests, such developments are going to have profound environmental impacts, both at the local and global level. Modernity, pursued at such velocity and on an historically unprecedented scale, is delivering a multitude of socio-cultural transformations that run counter to our current notions of sustainability. This chapter focuses on one aspect of that transformation: the household. But rather than rehearse the arguments of Myers and Kent's analysis we want to head in a more qualitative direction and explore transformations in housing and households in the urban environment of Phnom Penh, Cambodia's capital. Whilst Cambodia lies outside their category of 'new consumer countries', there are many 'new consumers' in Phnom Penh, offering an extremely vivid picture of wider patterns in the developing world around discourses and practices of sustainability. Cambodia, and in particular Phnom Penh, offer us important insights for a number of key reasons. Firstly, with the country departing from an era of sustained conflict and genocide, there is a discernible and tangible acceleration towards modernity. As one of Asia's poorest countries, economic mobility at the individual and family level often means being 'lifted out of' some of the lowest categories of poverty, as defined by developmental agencies like the World Bank, Oxfam and so forth. Secondly, and as we shall see, Cambodia offers a rich example of how such development organisations have influenced the parameters and discursive boundaries of what constitutes sustainability and sustainable development. Thirdly, Phnom Penh tells us much about how the aspirations of modernity in developing countries come to be played out in and through the home, and how such processes often stand outside the discussions around 'sustainability' that prevail in such national contexts.

Framing Sustainability in Cambodia

Over the last two decades Cambodia has received a vast amount of international attention and financial aid. Much of this has revolved around post-conflict recovery and the rebuilding, reconstruction or revival of the country's social, physical, political, legal, cultural and institutional infrastructure. In the early 1990s the country witnessed the arrival of more than 40,000 expatriates for what was, at that time, the biggest United Nations peacekeeping operation ever. Around that time governmental and non-governmental humanitarian aid agencies established themselves in Phnom Penh, forming the largest single sector of the Cambodian economy. As the decade progressed, the Royal Government and the Cambodian population would come to be heavily dependent on foreign aid and all its associated paraphernalia. More recently however, with increasing regional investment, Cambodia's capitalist economy has eclipsed the aid-dominated economy of the 1990s. Nonetheless, many aid organisations remain – even as late as 2009 the Cambodian Rehabilitation and Development Board Council for the Development of Cambodia (2009) lists almost three hundred separate non-government organisations actively working in the country, the vast majority of which are described as foreign. It is also worth noting this is a conservative total, as the Board lists more than 900 others – predominantly smaller local non-government organisations, but also a considerable number of foreign non-government organisations – which may still be active, as they have not reported their status to the board.

Not surprisingly then, for nearly two decades international aid agencies have been instrumental in mapping out a discourse of sustainable development, a language and paradigm familiar to the development sector as a whole and to many developing countries across the world (see also Asian Development Bank 2010). In brief, with development focused on poverty reduction, the term 'sustainable' invariably prioritised the need for long-term, stable, socio-economic growth. Although such a paradigm operated across multiple sectors and scales, the idea of 'sustained' poverty reduction became a particularly popular mantra for programs operating at the 'community' level. For a government facing a multitude of profound challenges, sustainability was understandably defined primarily in economic terms, with key documents such as the two national 'Socio-economic Development Plans' produced for the periods 1996–2000 and 2001–2005 adopting such terminology to orient the country's recovery (Ministry of Planning 1996, 2000).

This does not, however, mean environmental issues have been ignored entirely. Indeed throughout the conflict recovery period numerous non-government agencies established projects addressing environmental degradation. The Tonle Sap, known as Cambodia's Great Lake, for example, became the focal point of numerous environmental projects. A vast inland wetland, the Tonle Sap is a vital refuge for protecting some of Asia's most globally significant biodiversity. With an annual expansion and contraction ranging from 2,500 km² to 16,000 km², the lake is also an immensely important resource for fresh water and fish, providing 40–70 per cent of the protein intake of Cambodia's population at any given time (www.tsbr-ed.org/english/default.asp).Nominated as a biosphere reserve under UNESCO's Man and the Biosphere Program in 1997, the Tonle Sap has been the location of countless conservation and community-based projects, with organisations as diverse as the World Wildlife Fund and the World Conservation Union, the Enfants du Mekong, Mines Advisory Group and Buddhism and Democracy all establishing projects in and around the lake. In fact in 2010 the Tonle Sap Biosphere Reserve Environmental Information Database listed a staggering 782 projects operated by non-government agencies (Tonle Sap Biosphere Reserve 2010).

For many organisations like the United Nations Development Program (UNDP), conservation projects have often been folded into wider developmental frameworks, as evidenced in the Tonle Sap Sustainable Livelihoods and Chong Kneas Environmental Improvement Project initiatives, both of which combined the management of fisheries, tourism development and schemes fostering community-based protection of the fragile eco-reserve. In such cases we can clearly see the discursive structures of 'sustainable development' at play, whereby models of economic development that do not undermine the environmental resources they rely upon are advocated and advanced.

'Cambodia exports wooden logs and imports wooden toothpicks'[2]

Cambodia's forests provide a parallel example to the case of the Tonle Sap. Over the last forty years or so vast areas of the country's forestry reserves have been logged. The scale and speed of logging greatly increased in the 1990s, largely due to increased cross-border trade and associated forms of corruption. Bodies like the Asian Development Bank (ADB) and the World Bank have worked extensively with the Royal Government to try to improve the legal protection of forestry reserves and national structures of concession management for these regions. The following excerpt from the ADB's Country Assistance Plan 2001–2003 lays out the concerns and goals pertinent to this issue:

> If deforestation is not slowed, the resulting reduction in crop yields because of soil erosion and increased flooding may contribute to poverty in a rural population that is heavily dependent on agriculture. A related problem is that fisheries, a vital source of protein in the diet, are threatened by the combined effects of habitat loss in the inundation zone of the Tonle Sap, overexploitation, destructive fishing practices, inadequate sanitation, and agrochemical pollution. Improved natural resource management is a critical litmus test of the Government's resolve to reform and improve the lives of average citizens. (ADB 2000: 12)

What we begin to see here then is not only the ways in which notions of sustainability have been constructed in Cambodia, but also the objects upon which such discourses have focused. Environmental concerns have primarily focused on the degradation of so-called 'natural resources'. Nature has been construed as something external, ontologically separate from humanity or, to be more precise, 'human development'. Human involvement is only considered when it directly affects these resources. The discourse of this nature 'conservation', which emerged from the early 1990s onwards, has been overwhelmingly rural in its focus, where rivers, forests, birds, tigers and other wildlife were under threat. Where 'environmental impact' concerns have been discussed in urban contexts like Phnom Penh, the primary issue at hand has been the deterioration of residents' living standards through poor air quality, inadequate sewage treatment and ever growing traffic levels. In the case of forestry, the focus is entirely on the practice of logging, geographically situated in the forests, as well as on those who profit directly from this practice. Little to no attention is given to the consumption of

---

2   A contemporary Cambodian saying. Translation by Hem Tola and Willem Paling.

the products produced from this illegally logged hardwood, as they are made into furniture and sold in urban retail outlets.

Locating Climate Change Elsewhere

In essence, then, within the paradigms of sustainability and sustainable development that were prevalent in Cambodia over the 1990s and 2000s, the elements of 'nature' deemed under threat have been those that are both visible and empirically verifiable at the local or national level. The specific points of focus of the countless 'environmental impact' studies, undertaken by government and non-government bodies, conducted for proposed or existing developmental projects, exemplify this point. More recently though such environmental discourses, as articulated at the governmental level, have been extended to include more global concerns. Recent climate change initiatives developed by Royal Government include the Cambodian Climate Change Alliance (UNDP 2010), a program that replays the now familiar partnerships between government departments, international agencies and internationally funded non-government organisations.[3]

Not surprisingly, this has also meant the introduction of an abstract rhetoric; in this case 'climate change' has become embedded within, and thus co-opted by, a whole series of divergent, localised political agendas. As observed more broadly by Hughes (2009), the practice of local actors manoeuvring to capture aid and resources occurs against a general resentment of the hierarchies of power in the development sector. The result is an ambivalent response that 'while authorities may cooperate in the short-to-medium term with donors to increase political legitimacy, in the absence of political will, original patterns re-emerge' (Hughes 2009: 153).

Indeed, a number of recent events have seen the Cambodian government use 'climate change' as an excuse for problems seen by non-government organisations and/or the Cambodian public as being caused by actors with strong ties to government. Notable events in 2009 and early 2010 include the attribution of low water levels in the Mekong river to climate change, with many non-government organisations linking this to Chinese dams upstream (Sambath 2010); the death of 54 tonnes of farmed fish, attributed to rising water temperatures due to climate change, with farmers laying the blame on toxic waste coming from a South Korean-owned bio-ethanol factory (which had caused the death of 60 tonnes of farmed fish in August 2009); and a fire in a run-down area of Tuol Kork district that appears to be earmarked for development, in reference to which Cambodian Prime Minister

---

3   Investment in such initiatives by developed countries occur under the influence of economic motivators such as the Clean Development Mechanism (see UNFCCC 2006), a part of the Kyoto Protocol through which developed nations are able to earn tradeable emission reduction credits if they implement an emission reduction project in a developing nation.

Hun Sen suggested that such occurrences will become more common as a result of climate change.

It is entirely understandable that the Cambodian government locates the source of climate change as being elsewhere in the world. In 2006 Cambodians produced of 0.05 tonnes of carbon dioxide per capita from the burning of fossil fuels, in contrast to 20.58 tonnes in Australia, 31.41 tonnes in Singapore and 19.78 in the United States. With the increase in motor vehicle use in Cambodia, this figure would be higher if data were available for 2009, but still a far cry from the levels of 'developed' nations. What is particularly interesting, however, is that these recent events indicate that as they locate the source of climate change elsewhere, the government appears to be willing to use the abstract global concept of climate change as a scapegoat for apparently local problems.

Modernity and Consumption of Goods

Although Cambodia continues to be defined as one of Asia's Least Developed Countries, or LDCs, this term belies the rapidly shifting urban culture in Phnom Penh and elsewhere. Over the last ten to fifteen years, as Cambodia has embraced the multitude of connectivities and flows familiar in contemporary globalisation, its urban centres have increasingly become defined and characterised by large-scale consumption. To use the terminology of Myers and Kent, Phnom Penh has in recent years, seen huge increases in its population of 'new consumers'. The number of cars imported into the country each year has continued to rise rapidly, as has the retail space selling imported furniture, domestic appliances and electronic goods. The speed of change in Phnom Penh's retail environment has been one of the key markers of Cambodia's recovery and emergent, new economy. As recently as 2002 much excitement surrounded the opening of a new shopping mall as it would be home to the country's first electronic public escalator. Guides and signboards were introduced to help shoppers safely negotiate this new, unfamiliar technology (see Figure 4.1).

Cambodia is rapidly following a trajectory of increased consumption and energy use. Rising levels of wealth, which have come about since first Cambodia embraced neo-liberal capitalism in the early 1990s, have resulted in increased numbers of home appliances, cars, motorbikes and air-conditioning units. In recent years, brands like Adidas, Nike, Sharp, Volvo, Toyota, Lexus, Akira and Sony have all established franchises and showrooms in the country. The arrival of service centres for consumer electronics and other goods also enabled the concept of a national warranty to be introduced.[4] Together these were major departures for a country better known for a retail economy – most notably in the internationally

---

4   Previously, to obtain a warranty for items like laptops, cameras and other high-end electronic products required a trip to Bangkok, Kuala Lumpur or Singapore.

Figure 4.1   Shopping mall escalator 2003 (Photo by Tim Winter)

branded clothing and consumer electronics sectors – that largely centred around 'fake' goods and indirect grey imports.

Today, the city of Phnom Penh has a vastly different appearance to that of the 1990s. Roads are sealed, international fast-food chains are popping up, and the young drive to Phnom Penh's many universities on current model Japanese motorcycles, speaking on the latest mobile phones. For many of those with less financial capacity, Vietnamese and Chinese motorcycles and motorised bicycles – not readily available in the 1990s – are within their reach. It is a thriving and rapidly changing city, where many modern aspirations are realised. For those left behind, desires for these commodities are increased by their new prominence, with the increasing numbers of newly wealthy – new consumers – regarded with both awe and resentment. The response to these developments within Cambodia is mainly positive. Widespread aspirations for modernity involve increased use of technology and greater engagement with all forms of mobility, with the realisation of such aspirations understandably celebrated. Among particular groups of the newly rich, it is the ability to consume that signals modern status, and one's distinction proclaimed. Further, following this trajectory of rising modern consumption practices involves the pointed rejection of the practices of the lower classes, many of which are very environmentally sustainable.

Modernity, Architecture and Air-conditioning

Against such developments and cultural changes, it is easy to forget that the majority of Cambodians still live in relatively 'traditional' dwellings: wooden or bamboo houses raised on stilts, with gaps in the flooring allowing the cooler air beneath the house to be drawn into the living area as the hotter air escapes. Urban areas are dominated by Chinese-style shop-houses, generally well ventilated, with tiled floors and high ceilings. The cheap, hollow bricks typically used in the construction of these dwellings act as quite effective insulation. The residential housing of the masses continues to implement a practical heritage of cooling. However, this practical knowledge found in the houses of the 'lower classes' is rarely celebrated as heritage. Aspirations to modernity and upward social mobility specifically involve dissociation from this pre-modern vernacular architecture and associated methods of heat management. The use of air-conditioning is perceived as a superior, more desirable method of cooling indoor spaces. Such examples of Cambodian tradition, like those of many Asian cultural pasts, are treated as artefacts, inspiring the current generation in terms of historical greatness and power (London, Pieris and Bingham-Hall 2004). In contrast, the imaginings of an architectural future are predominantly modern in aspiration.

Following Cambodia's independence in 1953, former king, Norodom Sihanouk, began to build his vision of a modern nation. It was as part of this vision that a particular style of architecture, coined 'New Khmer Architecture', arose (Grant Ross and Collins 2006). Conceived at a time when air-conditioning was prohibitively expensive to install and run, this predominantly public and commercial architecture was designed to facilitate airflow and natural cooling to counteract the tropical heat. The chief architect of this movement, Vann Molyvann, undertook carefully planned, well-considered research, referencing the various Cambodian urban centres of the past 2000 years (Molyvann 2004). The heritage of these long-lasting, structured societies was one of the reference points for the development of a style of architecture and planning that would form the physical environment of new urban centres intended to be of comparable greatness.

Cooling features such as the iconic fanned concrete roof tops and double-brick walls shielded Molyvann's buildings from the tropical heat. Many buildings were raised on stilts, consistent with Cambodia's pre-modern vernacular architecture. Windows were positioned in order to avoid the path of the sun. The use of stilts demonstrates Molyvann's regard for the practical heritage of cooling found in the Cambodian vernacular. These implementations of local knowledge were combined with innovations enabled by modern scientific research – techniques gained by Molyvann, along with a modern aesthetic, while studying under the pioneering modern architect, Le Corbusier in Paris.

In their appraisal of New Khmer Architecture, Helen Grant Ross and Darryl Collins are critical of 'the pretentious monstrosities of Phnom Penh's present new rich' (2006: 133), noting that 'these buildings are all air conditioned' (2006: 136). Similarly, Molyvann himself, who served as Minister of Culture, Fine Arts, Town and Country Planning from 1993 to 2001, is often critical of projects undertaken by the current generation. He cites the greatest difficulties of his tenure as being 'lawless speculators who attempted to contort or blatantly disobey rules', likening the situation to 'a state of anarchy' (Molyvann 2004: 229). Through the mainstream Cambodian press, the current government directs retaliatory criticism at Molyvann. In response to his strong criticism of the filling of an urban lake, Pa Socheatavong, Phnom Penh's deputy governor told Molyvann that he 'should catch up with the times, that the ideas by older generation people are not always right, and that the younger generations are not all ignorant' (Heng 2008). Many of the sustainable cooling features of Molyvann's buildings have been modified and subsequently rendered ineffective. In one example raised floors were filled in to create new residential and commercial spaces. Speaking of the Vietnamese architect who modified the building in the early 1990s, Molyvann lamented that 'he walled in the balconies, although they were meant to provide some coolness', adding that 'they spend fortunes on air conditioning whereas the building was initially designed so it does not need it!' (Gée 2008). Molyvann's criticism is expressed in economic terms, outside the discourse of environmental sustainability, but it could easily be positioned within it.

Phnom Penh's recent construction boom has been driven by strong economic growth and increasing foreign investment in the lead-up to 2008. This boom has been accompanied by a trend in upward social mobility. Increase in wealth results in increased numbers of air-conditioning units, as well as the number of houses and other buildings that depend on air-conditioning. This newer housing is generally built in forms common throughout modern, urban South-East Asia. It indicates a desire to be a part of a global or regional modernity that references modernity elsewhere, rather than Cambodia's own early modern or pre-modern heritage. It is only in a handful of the most expensive new construction projects that natural cooling is considered a benefit. Two articles from the Cambodian magazine *Elegant Homes* (Frommer 2008, Edminston 2008), illustrate two very different views of heat management with respect to modernity, heritage and cooling practices. Jessice Frommer (2008) focuses on Angkor International Airport, designed by a French multi-national firm and which references from Cambodia's Angkorean heritage, as well as the more recent heritage of New Khmer Architecture. Jordyan Edminston (2008) focuses on a project run by a Korean firm, demonstrating a modern approach to cooling. Links between cooling and heritage, however, are not found in the company's marketing material.

Aesthetically speaking, the design of Angkor International Airport at Siem Reap by the French-run Archetype Group is a modern interpretation of traditional Khmer forms. The carved finials that adorn the outer apexes of many traditional buildings are replicated in simplified geometrical forms. Looking through Archetype's Cambodian portfolio,[5] it appears that this particular style – referencing local culture and ancient heritage (or constructions thereof) – is used primarily in hotels

---

5   Archetype has offices in Thailand, Dubai, India, Cambodia and France.

and other tourist-oriented sites. Such a display can be viewed as self-orientalising cultural performance on the part of the Cambodians who commissioned Angkor International Airport. They produce a construction of their own culture that caters to the imaginings and desires of foreign tourists. Such desires include the heritage of sustainable cooling techniques of a pre-air-conditioning era. Archetype Cambodia's Swedish co-owner, Bernard Wouters, praises the architecture of the Sangkum period: 'Like most people, I am very much interested in and admire the works of famous Cambodian architects such as Vann Molyvann'; this is followed by the author's statement that 'in the past, Cambodian architects were aware of using positioning and materials to ensure maximum wind-flow and shade to keep homes cool during the hottest months' (Frommer 2008: 47). Implicit in this statement is the observation that such knowledge is a property of the past, not of the present, but that it is a very worthwhile body of knowledge that Archetype is usefully reviving.

In contrast, the Korean developers behind De Castle promote their building as having 'been design[ed] by international experts to maximize the use of natural light and breeze for the most pleasant and comfortable living experience as well as ensuring fabulous city views' (De Castle 2009). Rather than drawing on tradition, cooling in these apartments, both in terms of making use of natural conditions and modifying natural conditions, is communicated in terms of a modern, scientific approach. De Castle claims to have been designed according to a well-researched understanding of Cambodian culture. Here a very different understanding of culture – as a property of contemporary Cambodia – results in a modern building that bears no resemblance to dominant understandings of Cambodian culture and heritage. Its depiction as the only high-rise in the city – a common feature of the marketing of all similar buildings, such as Gold Tower 42 – plays into desires for modern class distinction. De Castle is one of over 40 buildings of 15 floors or more that are currently under construction in Phnom Penh (see Figure 4.2). By comparison, in July 2010 there were just two completed buildings of 15 floors or more.

Being relatively removed from pre-modern forms of existence, there is a nostalgic tendency in more developed societies to romanticise rural poverty and idealise the pre-modern. Such an approach has been observed as a major problem for architects who draw on more traditional Malay architecture in contemporary projects. 'Class and cultural boundaries between the urban elites and the rural or tribal sources of vernacular architecture often convert its appropriation into a predatory act with colonial underpinnings' (London, Pieris and Bingham-Hall 2004: 21). Drawing in this manner on the aesthetic associated with the lower classes contrasts with the modern aspirations found throughout Asia. While the analysis of a much greater number of projects is necessary to deduce any sort of trend, this brief analysis shows that at least in these few cases, Westerners promote their tourist-oriented design as significantly influenced by Cambodian traditions and implement 'green' measures as part of the design. This is linked to their background in and desire to appeal to an understanding of sustainability

Figure 4.2   High-rise construction in Phnom Penh 2010 (Photography by Willem Paling)

and consumption that is predominantly located in the global North or West. In contrast the Asian-run projects promote their Cambodian-oriented design as 'international', that is, anything but Cambodian. 'Green' concepts are not featured in their marketing materials.

This positioning provides the distinction of being situated within a modernity that is larger than Cambodia, confirmed by its many references to the 'international'. The public response to these developments is one of general excitement, with advertising for developments such as CamKo City, Gold Tower 42 and the International Financial Center evoking images of a Phnom Penh in the near future that fits with mainstream imaginings of what a modern global city ought to look like. These developments occur outside the non-government organisation-dominated environmental sustainability discourse as they are developed and funded by private enterprise, most of which comes from South Korea. Critical voices have little influence and are relatively few in number. Even so, this criticism tends to be framed in economic terms, for example, recent criticism has been levelled at construction delays stemming from the 2008 financial crisis, or made in terms of claims of government graft stemming from foreign investment. The general trajectory of Phnom Penh, towards a modern city with widespread high levels of household consumption, is not subject to any popular criticism.

Waste Disposal and Recycling Networks

At the other end of the spectrum, very sustainable household recycling practices are taking place, enabled by poverty. At rural weddings young children crowd around the fences gesturing for empty drink cans from the wedding guests. The cans are then sold to a local scrap dealer. It seems children from middle class families have no interest in the cans so actively sought by rural children. This may in part be due to them being given small sums of money by their families, but it is also because they view the practice as undesirable – associated with the lowly social status of those poor enough to need to gather other people's rubbish. In almost any populated area of Cambodia one can confidently throw a can on the ground knowing that it will soon be picked up and be on the path to being recycled. In urban areas predominantly ethnic Vietnamese recycling collectors walk the streets calling for householders to bring out their cans, bottles and other recyclables, which are subsequently weighed and purchased for a small sum (see Figure 4.3). For some middle class households that do not need the small change, the fact that these collectors fossick through their garbage for recyclables enforces the distinction between recyclable and non-recyclable waste. For the recyclables that manage to make it into landfill, scavengers at the dump are many, mining the city's waste for anything that is economically viable to recycle.

As reported in the *Cambodia Daily* in April 2009 (Marks and Reuy 2009), in Phnom Penh these networks of recyclers consist of the collectors noted above who roam the streets, and scavengers at the Stung Meanchey dump who on-sell

Figure 4.3   A Recycling Collector Pulls his Cart after Sifting through Roadside Waste in the Garment Factory District of Chom Chao, Phnom Penh (Photo by Willem Paling)

their collected recyclables to middlemen, who sell them on to larger collectors and ultimately to exporters, who send them to Thailand or Vietnam. The price of these goods is affected by regional political issues, such as the border dispute between Cambodia and Thailand, as well as global economic issues, with global demand for aluminium directly linked to the price of used cans, and the price of oil linked to the price of recyclable plastics. The combination of an unpleasant work environment, low income opportunity, low social status, and the very real threat of disease associated with working as a waste collector (Xinhua 2006) mean that it is not in any way a desirable occupation. It is made possible by the very large number of people willing to work for very small financial reward (Rathana 2009), a situation that plays into a widespread and longstanding contempt for ethnic Vietnamese in Cambodia. It is reasonable to expect that if alternative employment were to be made available for these workers, this disorderly, but comprehensive network of recycling would suffer.

The recent opening of a new dump in Phnom Penh has brought about a conflict over the right to pick through rubbish and to purchase recyclables from scavengers. According to the *Phnom Penh Post*, officials managing the dump have demanded payment for the right to continue operations at the new site (Sophakchakrya 2010). As the profit from this recycling is spread further it is reasonable to expect that the efficiency and thoroughness of the recycling process will suffer. Driven as it is by economic pragmatism, this recycling process could not be sustained if it were to become unprofitable. Similarly, economic pragmatism drives the general awareness of fuel efficiency of different models of motorbikes for the middle classes. Brands of motor oil are advertised using claims of fuel efficiency, although this refers to economic benefit rather than reduced environmental impacts. For businesses and non-government organisations proffering sustainable goods, marketing success tends to be more common through focusing on the short-term economic benefits rather than longer term benefits, or the more abstract concept of global climate change (A. Ahmed personal communication 23 March 2010). For example Ahmed has had success in marketing a product that extends the life of the diesel generator-charged car batteries that power many rural households, a small investment that sees immediate returns. However he has had little success in marketing solar systems that would see significant savings after six months.

Conclusion

When viewed in the context of contemporary Western understandings of sustainability, many Cambodian household practices are exemplary. In both rural and urban Cambodian households there is a tendency towards very low levels of energy use, low levels of consumption of non-renewable resources and high rates of recycling. However these practices are under threat – from the increases in wealth and corresponding decline in poverty – meaning that their economic utility is diminishing. But as we have seen, many of the practices involved in household construction and consumption in Cambodia are rapidly changing as the project of economic development experiences visible and significant success. It is difficult to reconcile the seemingly contradictory concepts of environmental sustainability and economic development, the real-world models for a 'developed' society and particularly those to which Cambodians aspire, all being grossly unsustainable. When seen in a wider context it is readily apparent that such processes are far from unique to Cambodia. As we have argued, the situation there offers valuable insights into many of the changes and challenges now unfolding in developing world contexts; the majority of which have yet to receive the critical attention they deserve.

References

ADB, *see* Asian Development Bank
Asian Development Bank. 2000. *Country Assistance Plan 2001–2003*. [Online]. Available at: www.adb.org/Documents/CAPs/cam.pdf [accessed: 7 September 2010].

Asian Development Bank. 2010. *Country Assistance Plans, Cambodia, Country Performance Assessment*. [Online]. Available at: www.adb.org/documents/ CAPs/CAM/0103.asp [accessed: 7 September 2010].

Beck, U. 1992. *Risk Society: Towards a New Modernity*. London: Sage.

Beck, U., Giddens, A. and Lash, S. 1994. *Reflexive Modernization: Politics, Tradition, and Aesthetics in the Modern Social Order*. Stanford: Stanford University Press.

Bennett, T. and Healy, C. 2009. Assembling culture. *Journal of Cultural Economy*, 2(1–2), 3–10.

Cambodian Rehabilitation and Development Board Council for the Development of Cambodia. 2009. The Cambodian NGO Database. [Online]. Available at: http://cdc.khmer.biz/ngo/report/listing_by_lastupdate.asp [accessed: 7 September 2010].

Connolly, J. and Prothero, A. 2008. Green consumption: life-politics, risk and contradictions. *Journal of Consumer Culture*, 8(1), 117–145.

De Castle. 2009. *Site Design*. [Online]. Available at: www.decastle.net/design/ site_design.html [accessed: 13 September 2010].

Edminston, J. 2008. De Castle – bringing luxury apartments to Cambodia. *Elegant Homes*, 1(2), 80–89.

Frommer, J. 2008. Archetype Cambodia – combining past and present to create innovative design. *Elegant Homes*, 1(2), 40–49.

Gée, S. 2008. Lives in the Building (1): Once upon a time, two Buildings in Phnom Penh. *Ka-set* [Online]. Available at: http://cambodia.ka-set.info/culture-and-society/news-building-phnom-penh-urbanism-architecture-080530.html [accessed: 17 August 2009].

Giddens, A. 1991. *Modernity and Self-identity: Self and Society in the Late Modern Age*. Stanford: Stanford University Press.

Grant Ross, H. and Collins, D. 2006. *Building Cambodia: 'New Khmer Architecture' 1953–1970*, Bangkok: Key Publisher.

Heng, S. 2008. Pa Socheatvong to Vann Molyvann: Don't look down on the younger generation, we are not all ignorant … we know how to fill our pocket better than you. *KI Media* [Online]. Available at: http://ki-media.blogspot. com/2008/01/pa-socheatvong-to-vann-molyvann-dont.html [accessed: 17 August 2009].

Hughes, C. 2009. *Dependent Communities*. New York: Cornell University Press.

London, G., Pieris, A. and Bingham-Hall, P. 2004. *Houses for the 21st Century*. North Clarendon VT: Tuttle Publishing.

Marks, S. and Reuy, R. 2009. Scavengers seek framework for waste recycling. *Cambodia Daily* [Online]. Available at: www.camnet.com.kh/cambodia.daily/ selected_features/cd-Apr-15-2009.htm [accessed: 29 March 2010].

Ministry of Planning. 1996. *First Socioeconomic Development Plan 1996–2000*. Phnom Penh: Ministry of Planning, Royal Government of Cambodia.

Ministry of Planning. 2000. *Second Socioeconomic Development Plan 2001–2005*. Phnom Penh: Ministry of Planning, Royal Government of Cambodia.

Molyvann, V. 2004. *Modern Khmer Cities*. Phnom Penh: Reyum Publishing.

Myers, N. and Kent, J. 2004, *The New Consumers: The Influence of Affluence on the Environment*. Washington: Island Press.

Rathana, K. 2009. *Solid Waste Management in Cambodia*. Phnom Penh: Cambodian Institute for Cooperation and Peace.

Sambath, T. 2010. Low Mekong isn't caused by dams: govt. *The Phnom Penh Post* [Online]. Available at: www.phnompenhpost.com/index.php/2010032934421/National-news/low-mekong-isnt-caused-by-dams-govt.html [accessed 29 March 2010].

Sophakchakrya, K. 2010. Refuse Site: scavengers want access to dump. *The Phnom Penh Post* [Online]. Available at: www.phnompenhpost.com/index.php/2010032434181/National-news/refuse-site-scavengers-want-access-to-dump.html [accessed: 30 March 2010].

Tonle Sap Biosphere Reserve. 2010. Environmental Information Database, NGO Network. [Online]. Available at: www.tsbr-ed.org/english/ngo_network/ngo_networks.asp?province_id [accessed: 7 September 2010].

UNDP, *see* United Nations Development Program

United Nations Development Program. 2010. *Launch of Cambodia Climate Change Alliance*. [Online]. Available at: www.un.org.kh/undp/Climate-Change/Launch-of-Cambodia-Climate-Change-Alliance.html [accessed: 23 March 2010].

UNFCCC, *see* United Nations Framework Convention on Climate Change

United Nations Framework Convention on Climate Change. 2006. *CDM: About CDM. Clean Development Mechanism*. [Online]. Available at: http://cdm.unfccc.int/about/index.html [accessed: 23 March 2010].

Xinhua. 2006. *Cambodia: 14 Diseases Caused By Rubbish, Collectors at Risk – Health News – redOrbit*. [Online]. Available at: www.redorbit.com/news/health/399688/cambodia_14_diseases_caused_by_rubbish_collectors_at_risk/ [accessed: 29 March 2010].

Discussion: Interrogating the Household as a Field of Sustainability

Gay Hawkins

These chapters offer rich insights into the cultural and political complexities of the household as a site of analysis. They provide fascinating accounts of how 'sustainability' acquires political and practical meanings in places as diverse as a rapidly developing Asian city: Phnom Penh; a major regional centre: Wollongong south of Sydney; and on the national scale: Australia. They also pose a range of conceptual, methodological and political questions about how to critically evaluate the role of households in progressive environmental change. In seeking to comment on their central themes and concerns three issues emerge: firstly, the framing category of the household and the possibilities and limitations it presents as a mode of analysis; secondly, how to investigate 'household cultures' and the ways in which culture is materialised; and, finally, the politics of households: how do they engage with political processes and how do these engagements mediate the larger political dynamics of governmentality or activism? I shall consider each of these in turn.

There is no question that 'the household' is a central focus in environmental policy and a key site of governmental intervention. But, as these three chapters reveal, 'the household' is a tricky category to pin down. For Aidan Davison the idea of the household is deeply connected to the history and ideologies of home ownership and home-centredness in Australia. This challenges the tendency to define households using abstract quantifiable measures. While statistics about households might be central to the representation of census data or economic growth, this technique does not capture the complex cultures of domesticity and identity that senses of home generate. In a similar vein Chris Gibson and colleagues refer to 'household cultures' to emphasise the ways in which bodies, ideologies, technologies and materials all work to constitute a space called 'home'. Both these chapters show that the household invoked in environmental policy is highly normalised and constituted through specific empirical processes. In contrast to this, 'home' emerges as a complex spatial and temporal field where everyday life unfolds. For those able to inhabit a home, and this is by no means assured for very poor residents living in Phnom Penh, for example, a sense of ontological security is often produced from the connection between being and dwelling.

The value of this focus on the household as central to the meaning of everyday life is that it foregrounds the ways in which home is often experienced as a space of containment and security. Aidan Davison refers to research on Australians' recent obsession with renovation to show how changing and modifying domestic space generates a sense of control and order In an increasingly chaotic and uncertain world. Chris Gibson and colleagues explore the ways in which normative expectations of sustainable domestic practices from government or the media are negotiated in the material circumstances of situated households, with their very specific habits and expectations about ways to live. Willem Paling and Tim Winter show how pre-modern vernacular urban design practices in Phnom Penh produce relatively cool homes in a very hot climate through the use of simple and sustainable technologies from gaps in the floor to high ceilings. In each of these examples household practices serve to secure well-being. However this ontology is always contingent and unstable because homes are not contained; they are not isolated enclaves. They are nodes in complex and multiple networks of infrastructure, flows of money and commodities, people, media, matter and more.

Households are contained and networked at the same time, and managing the tensions between this forcefield of internal and external relations is at the heart of sustainability. If a sense of ontological security comes from controlling domestic space as the one site where people feel relatively free to 'do their own thing', how then is the impact of wider social and environmental change, or governmental policy shifts negotiated? Each of these chapters examines the dynamics of change in households and raises the fundamental question of how ways of living are transformed. Too often policy targets the household as an easy option, as a zone where changes in everyday life are possible and relatively straightforward. This strategy often ignores the complexities of household practices and the ways in which the dynamics of urban assemblages and governance reverberate on homes, actively inhibiting progressive change. For example, not wanting to drive in the interests of reducing one's carbon footprint is totally dependent on having a home that is well connected to public transport. On the other hand, changes in the infrastructure servicing and connecting homes can prompt sudden and effective transformations in everyday life and household practices. The reduction of waste removal services and introduction of compulsory household sorting and recycling in most Australian homes over the last thirty years has transformed how people relate to their rubbish. While some householders complained of state coercion when these policies were introduced, outraged that they were no longer free to make as much waste as they liked, most accepted this structural shift by reducing what they threw out and how they actually did it. In this way many householders felt connected to wider environmental issues beyond the home both ethically and practically. In putting out the garbage properly they were doing their bit for the planet.

Households, then, are interfaces. Altering the ways in which they are connected to wider urban assemblages inevitably shifts habits and domestic cultures in complex ways. Each of these chapters offers excellent insights into how to analyse these interactive processes and how to think about the household not as a set of abstract quantifiable measures or as the outcome of wider structural logics but as a field of negotiation where everyday life is practised. The challenge is to assess how it is being constituted in different settings and in different relational networks.

These chapters rise to this challenge with their close focus on the ways in which technologies, consumption practices, habits, vernacular design and more generate a sense of home always in relation to the wider field of economic, cultural and political connections that households are caught up in.

The second theme that emerges is the materialisation of culture. All these chapters show how the dynamics of culture are realised through endlessly variable relations with the non-human. Culture, rather than being an expression of diverse human belief systems or values, emerges as a set of practices that involve multiple relations with this realm. This non-human stuff, from eco-shopping bags to air-conditioning to street rubbish, is not passive or inert in these relations. It participates in the constitution of cultural practices and meaning systems in ways that involve constant negotiations with material presences. In other words materiality matters and what Davison, Gibson and colleagues, and Paling and Winter show is how the non-human can be generative: inviting or suggesting different cultural practices. The other significant point that is developed is the way in which materiality is implicated in social change. Paling and Winter's account of recent transformations in the urban form of Phnom Penh is an incisive investigation of how different material forms participate in economic development. Their argument is not simply that sealed roads and shopping malls express rapid change but rather, that these material forms are helping to create a different or 'modern' ontological reality that interferes with other, more traditional and sustainable realities.

Finally these chapters invite us to extend our understanding of politics and political process. All suggest different ways of thinking about households as always open to the emergence of the political. Gibson and colleagues explore the ways in which ethico-political frameworks inform consumption practices. Rather than morality or environmental responsibility being seen as imposed from above and prompting guilt or grudging obedience, ethics are realised in practices – in the choices people make and the ways in which these choices are interrogated and justified.

Central to this approach is an implicit recognition of the generative tension between the political and politics. For Andrew Barry (2001) 'politics' are highly codified forms of contestation that rely on conventional devices and institutions such as government policies, environmental education campaigns and media debate, for example. In contrast, 'the political' is the unpredictable process of opening up new sites of dissent and contestation that may or may not adopt the logics of existing politics (2001: 207). This distinction between mainstream politics and the political resonates with Michel Callon, Pierre Lascoumes and Yannick Barthe's account of 'hybrid forums'. These are fields of dissent that emerge out of public interrogation and contestation of markets, commodities and their 'overflowings' or negative impacts. Michel Callon and colleagues argue that overflowings can generate political processes and innovative configurations with a range of actors, making clear distinctions between the institutions of markets, techno-sciences and publics difficult to ascertain. The issues and questions that proliferate in these configurations, and the reflexive activity they prompt, constitute spaces for the political (Callon, Lascoumes and Barthe 2009). This account of hybrid forums acknowledges the role of technical devices and procedures in enabling 'measured action' in response to dissent and interrogation of markets and commodities. However, they argue that these technical devices and procedures need to be dialogic, not just delegative, if they are to avoid limiting the spaces and possibilities for contestation and change, or what Barry calls the 'anti-political' effects of politics. Reactive opposition and adversarial techniques are often the least effective way to establish lasting and democratic collaboration. As Callon, Lascoumes and Barthe say:

> By trial and error and progressive reconfigurations of problems and identities, socio-technical controversies tend to bring about a common world that is not just habitable but also livable and living, not closed on itself but open to new explorations and learning processes. What is at stake ... is not only reacting but constructing. (Callon et al. 2009: 35).

This focus on the political as a process driven by generative and exploratory techniques, not just critique and opposition, is significant. It shows how markets, publics and matter can become caught up in relations of interrogation and dialogue, and how commodities are iterative processes, subject to continual qualification and re-qualification through household practices. It also shows how households can be crucial sites for experimentation with what kind of 'common world' we wish to create and inhabit. All these chapters foreground household cultures as spaces of possibility, connected and constrained and always susceptible to the emergence of the political.

References

Barry, A. 2001. *Political Machines.* London: Athlone.
Callon, M., Lascoumes, P. and Barthe, Y. 2009. *Acting in an Uncertain World: An Essay on Technical Democracy.* Massachusetts: MIT.

PART II
Domestic Spaces and Material Flows

Chapter 5

Beyond McMansions and Green Homes: Thinking Household Sustainability Through Materialities of Homeyness

Robyn Dowling and Emma Power

An increasing volume of research demonstrates the ways in which the possibilities and pitfalls of household sustainability are connected to materialities and imaginaries of home. Household energy and water use, for example, are undeniably informed by and constitute cultures of home, whether through cultures of comfort that underpin technology and energy use (Shove 2003), aesthetics and dispositions that shape water use in gardens (Askew and McGuirk 2004, Moran 2008), or the connections between cleanliness, class and water use (Sofoulis 2005). This chapter is positioned within this broad literature, and takes it in three new directions. Firstly, our conceptual entry point is that of materialities of 'homeyness'. Drawing on the early work of Grant McCracken and more recent work on materialities of home, we are interested in the practices, objects, consumption patterns and energy use associated with making houses homey: comforting, welcoming and, often, unpretentious. In the first section of this chapter, therefore, we outline the notion of homeyness and its possible connections with sustainable and unsustainable practices.

Secondly, we take the literature on home, households and sustainability as a focus for a site that is popularly considered to be the epitome of unsustainability: 'McMansions', or large, new dwellings located on the fringes of Sydney. In popular culture and some academic commentary, these houses are represented as over users of energy, and their inhabitants are cast as selfish and consumer driven (for example Curtin 2009). Surprisingly little empirical research has been conducted on household sustainability in Australia's outer suburbs, so we aim to partially remedy this gap here. Thirdly, rather than focus on the uptake of eco-efficient domestic technologies (cf. Hobson 2006) here we are explicitly interested in the ordinary rather than extraordinary, on the assumption that the everyday consists of a complex layering of sustainable and unsustainable practices. Hence we (re)read the ordinary in terms of sustainability: to what extent (if any) and in what form does sustainable practice take place within the fabric of everyday life carried out in suburban, single family dwellings? Our purpose here is to read 'against the grain', identify fissures in the ordinary and taken for granted and perhaps open up new possibilities for change. Thus in the second, empirical, section of the chapter we draw on interviews with residents of new suburban developments in Sydney to elaborate aspects of domestic practice that may illuminate household sustainability in unintended ways. In particular we focus on the intersections of familially connected objects; bodily feelings of spaciousness and flow that shape residents' feelings of home; and materialities of homeyness. We conclude by considering the importance of examining the ways that homeyness intersects with (un)sustainable practices, offering barriers to and opportunities for more sustainable modes of living.

Materialities of Homeyness

Creating a house that is homely, or homey, is central to the feeling of being at home in the house-as-home (Blunt and Dowling 2006). Homeyness is both a motivation for and outcome of homemaking practices as inhabitants materially and figuratively ensure that their house is a 'home'. Many of the imaginative referents of homeyness are both well known and well understood. These include feeling comfortable, secure and able to make independent choices (Blunt and Dowling 2006). Materialities of homeyness are equally important. Homeyness is a multi-sensory relation that grows through an embodied engagement with the spaces and objects of the house. As Hetherington (2003: 1939) notes in a related discussion,

> Whereas we enter our houses through the front door, we enter our homes through our slippers ... Our encounter with certain objects is more obviously tactile than it is visual. The feel of something can generate a sense of who we are and where we find ourselves – a sense of place – hence warm cosy slippers. But, of course, the door might assume this quality, too, especially if we have been away for a long time. The feel of the key in the latch, the click as it opens – or perhaps we need to nudge it in just the right place because it sticks a bit – are all familiar experiences that place us within the familiar. The place is not in the slippers or even in the sticking door but in an absence made present in what these experiences reveal to us – in this case the familiar, the routine, the ordinary, the known, through which we can recognise ourselves as particular subjects. These qualities are not represented by the slippers, but performed by them.

Feelings of homeyness grow through these embodied relations with the materiality of the house, including furniture, toys, food, colours, textures and even the house itself. They are multi-sensory and highly contextual, drawing in culture, practice and place.

The achievement of homeyness can entail very specific aesthetic and textural choices. For participants in Grant McCracken's North American study:

> 'Homey' colors are the 'warm' colors: orange, gold, green, brown. The preferred materials for interior walls are wood, stone, and brick. The only acceptable material for furniture construction is wood. Fabrics for furniture are relatively unfinished natural fibres. Fabric patterns are florals (especially chintz

or conversationals). Furniture styles are traditional, homemade, hand crafted, colonial or antique ... Objects are homey when they have a personal significance for the owner. (McCracken 1989: 169)

In these households careful combinations of the correct colours, textures and fabrics, mediated by culture, taste and personal significance, worked together to create a homey space that afforded a sense of comfort and familiarity to residents and their visitors. Objects that narrate social relations and familial connections also shape feelings of homeyness, an effect that is compounded by the dominant association between family and home. Gifted and inherited furnishings, for example, can symbolise familial connections and help to turn a house into a family home (Blunt and Dowling 2006, Noble 2004). The placement of objects and the practices that take place around them are equally significant. For example in co-habiting households the process of combining possessions and taking shared decisions around design and decoration is an important way in which houses are made into shared homes (Gorman-Murray 2006, see also Clarke 2002). Similarly, Gillian Rose (2003) shows that feelings of homeyness and familial connection emerge not simply from the content of family photographs but through the everyday practices that take place around photographs as material objects, such as sorting, sending and displaying them.

However homemaking is not a one-way process where people simply appropriate objects, furnishings, colours and textures to achieve feelings of homeyness. Rather it is a multi-directional relation where the materiality of the house also shapes and potentially surprises, disturbs and alters residents' sense of home. Daniel Miller (2002) describes this as a process of accommodation, where residents both appropriate objects and alter the ways in which they live home in response to the particular materiality of the house. In Miller's own home this is experienced as a 'haunting' whereby an undesirable colour scheme installed by previous owners challenges his sense of control over the domestic space by suggesting his own lack of taste to visitors. Nicky Gregson (2007) argues that these processes of change and accommodation are a central aspect of homemaking, pointing to the ways in which furnishing and decorative choices are shaped by the style and design of the house and existing furnishings. Robyn Dowling (2008) likewise foregrounds the ways in which residents of open-planned houses accommodate to the openness of the space, adapting their domestic practices to complement and sometimes circumvent the openness of the house design.

Feelings of homeyness are further shaped by factors that exceed human intentionality and design and that speak instead of a more-than-human dynamic that goes beyond the decisions and practices of current and previous residents, designers and house builders (Hitchings 2004). Houses are constructed in an environmental context; house designs that complement cycles of seasonal change and the diurnal rhythms of daylight and darkness can contribute to residents' sense of homeyness. Conversely, designs that do not harmonise with these rhythms can prompt changes to living patterns – for example avoiding particular sections of the house during winter – and give rise to feelings of discomfort and unhomeyness (Power 2009a). Ruptures in the borders that separate home from the outside world, such as the disruption of essential services like water and electricity (Kaika 2004), and unexpected incursions into the home by 'pest' animals (Power 2009b), can similarly disrupt feelings of homeyness and give rise to a competing sense of unhomeyness, of 'not being at home in one's own home' (Vidler, cited in Kaika 2004: 276). These challenges to residents' sense of homeyness prompt changes in homemaking practice, for example efforts to re-secure home against incursions.

In summary, the materiality of housing exerts significant effects on human residents. From carefully chosen furnishings and colour schemes to unexpected and undesirable happenings, the materiality of the house can shape and alter residents' sense of homeyness. These effects take place within a particular environmental, cultural and social context, and draw in factors ranging from house design to kinship ties and financial constraints. It is impossible to comprehend homemaking practice outside of these relations. With this starting point we move to introduce the empirical project on which this chapter is based.

Methods

Our focus is a study of the everyday practices and narratives of 26 Sydney households. In-depth interviews were conducted in 2004 as part of a larger project concerned with the aesthetics and material geographies of the mass-produced suburban housing that dominates the Australian suburban landscape. Interviews were conducted with residents of such houses. They were taped and various photographs of the home were also taken when permission was given. The interviews were transcribed and a thematic coding undertaken. Two-thirds of the houses were less than five years old, with an almost even divide between single-storey and double-storey houses. All but one had both a large open-plan living space (typically a kitchen and 'family room') and a separate and more formal living room. Although the advertisements seeking participants for this study were not gender specific, with few exceptions it was women who did the interviews. The women interviewed were living largely in middle-class nuclear families.

Adult members of all but four of the households were aged in their thirties and had two or three children currently living with them. Most of the children were under twelve. The majority were Australian born and there were no recent migrants to Australia. Most residents (both male and female) were in middle-class occupations, and all were owner-occupiers. The houses represented in the study were scattered across Sydney's outer suburbs principally because the recruitment strategies targeted those living in large-scale new-build developments where project homes predominate.

Household sustainability was not the central focus of this research (cf. Hobson 2006). Interviews focused on residents' perceptions and practices of home: their furniture provisioning, decorative practices, choice of housing style, size and layout, and use of different spaces within the house. This focus allows, nonetheless, an exploration of the ways in which everyday household practice may, inadvertently and in a contradictory manner perhaps, intersect with less resource intensive patterns of everyday life. In this chapter, the broad coding schemas around family, furniture and spaciousness are drawn upon. In particular, the first empirical section examines interrelations between furniture choice, kinship ties and reuse/recycling, while the second examines the performance of homeyness through the spaciousness and flow of the domestic space.

Homeyness, Family and Circulations of Furniture

Homeyness is materialised in the home in diverse ways, including, for example, through practices of decoration, as initially analysed by McCracken; and also smells, sounds, memories and textures. It became apparent in the interviews that homeyness was embodied in furniture, and in particular furniture with familial connections. Furniture provisioning was a key component of many of these householders' domestic practices. Numerous participants described their trips to famous furniture and department stores to acquire furniture for their new house: just the right couch to fit their family room, a new dining table to fill an empty space, new beds for children who had previously shared a room. Such newly purchased furniture, and the minimalist aesthetic it commonly represented, could certainly engender homeyness through its sense of order and flow, as we discuss below. But the acquisition of new furniture took place alongside the circulation of old furniture, and in particular its circulation across the generations. All households had at least one item of furniture that had been given to them or acquired from friends and family. Wall units, cabinets, bedding and chests were gifted by family; sometimes literally inherited. Such furniture also prompted feelings of homeyness, and the ways in which this occurred is our concern here.

Homeyness was associated with a sense that a space needs to look and feel 'lived in' – not messy, but certainly characterised by a mixture of objects and furniture – in opposition to the style found on television shows and in magazines. Rebecca was talking about how furniture in stores and magazines didn't 'feel homely'. When asked what she meant by this she replied: 'Just lived in so it looks like people actually live there and actually enjoy living there.' For Rebecca, this could be achieved through inherited furniture:

> *Well for example the lounge out here, we wanted it to look good and it had to be comfortable, but we knew it wouldn't be the sort of thing we'd be curled up on, so it didn't quite matter as much. The one we've got out there is just a really comfortable one; it was my husband's grandmother's. We want to get rid of it and get a corner lounge out there so we can stretch out, lay on it, lounge on it, but because we envisaged this as a bit more of a visitor place it's not so much to lie on the lounge.*

Figure 5.1   'Inherited' Furniture in a Formal Living Space

In this case the inherited lounge was ultimately to be replaced with a newer model that would better meet the ideals of a relaxed family room yet, in the meantime, the lounge could be comfortably accommodated within the family space where appearance was not as much of a concern as in the more formal 'visitor place' (see Figure 5.1).

Homeyness was also associated with an eclectic mix-and-match aesthetic which could be achieved through receiving gifts of furniture from others. For one woman, building and moving into her new project home involved a significant investment of time and money in new consumer goods. Nonetheless, this is her description of her living room:

> *A bit eclectic. Because as I said this was second hand, we didn't go looking for sofas, this came up at work and I went thank you. As you can see, again we've been given the buffet a friend didn't want so I went thank you. A friend of my parents didn't want the china cabinet so I went thank you. The piano my neighbour didn't want. I mean it has cost me a thousand dollars to get it fixed, but it's still second hand.*

Here aesthetic compromises enabled the creation of a homey space, one in which the family could live comfortably. As this woman says about this space, it is 'not trendy, just friendly and welcoming, friendly and welcoming so people can come in and make themselves a coffee and sit down'.

Figure 5.2   Homeyness and Inherited Objects

Spaces also needed to be relaxing for their inhabitants. Central to Figure 5.2 is a large wooden model-aeroplane that the inhabitants described as 'inherited': 'Some friends that went to England had that and we inherited it.' For the male householder, the plane was central to his sense of homeyness: 'It's just relaxing. I sit there, watch the TV and look up there.' In contrast, circulation via family did not always engender homeyness. Sarah had been given 'that buffet, a second-hand thing from someone who died when we got married. It's great to hold things but I hate the look of it'. Similarly, for Nicole the presence of grandma's 'corner thing' was to be tolerated rather than embraced:

> *Like this corner thing over there, that was my grandmother's. Don't really like that, but she's passed away now and my mother was clearing out her house and said grandma would love you to have that, so I didn't really want to say to Mum oh, it's really ugly, so I just put it over there and deal with it.*

For Nicole, then, the corner display shelf was retained and used because of its connection to family. Although not valued beyond this connection, its presence and functionality within the house replaced the need for purchasing a new item.

Finally, re-circulated furniture materially embodied one of the classic elements of homeyness: family. The placement and continued presence of such objects in these houses was dependent on familial relations. For Julie, her grandmother's furniture was deserving of a prime spot in the house:

> *And that box there, that was my grandma's. That was just in the corner in the family room before and I really wanted it. There wasn't a good place. We didn't have enough room in the last house to make it somewhere that would stand out more, and I really wanted it, because it was my grandma's too. I really wanted to put it so it would show up.*

The case of Julie illustrates practices of both keeping connected to family and the circulation of objects within the house. Julie had a granny flat built for her father when he came to live with them from England. He brought his own furniture with him. When he died, his furniture was moved into the main house, into the formal living area. Julie now described this as the 'English room', containing objects 'older than Australia'. She liked these objects: 'I find them warmer, yes warmer ... I love to look at them. Say with this bureau with its carving, I just like to feel it, and because I remember it from when I was a little girl ... Oh yes, oh yes, I love to have that furniture. I'd rather have that than go and buy it.' The granny flat became a temporary holding space for mobile children and their objects. The youngest daughter's furniture occupied the granny flat while she was away.

Family spaces were also frequently home to older items kept because they allowed a more casual style of living that allowed family members to relax and not worry about dirt and cleanliness. Justine kept her old couches for her dogs to sit on and also used them when she was eating a 'messy' meal, like spaghetti. Similarly Vera retained old couches for her children to use:

> *This is the games room. The kids play here, and with their friends. And if I've got extra things, furniture, I keep them here instead of putting them in the rubbish. They [couches] are old, very old, from the old, old house. But I just keep it for the kids, or just in case people are sometimes here.*

Children and children's spaces become the recipients of furniture that is no longer fashionable or is simply old. Rumpus and family rooms become dumping grounds for 'old stuff'.

Homeyness in these households had a number of intersecting materialities – of colours, aesthetics, textures and objects. In terms of furniture, which has been our focus here, furniture gifted by friends and family was associated with homeyness, and eclectic rather than modernist and minimalist aesthetics; while the parts of the house most identified as homey were family rooms, rumpus rooms and, sometimes, children's bedrooms.

Flows

While the previous section focused on the flow of furniture into and through the house, and the ways such flows were held together (or fell apart) according to notions of family and style, in this section we focus on people's senses of flow. Flow was both a visual and whole-of-body experience that gave rise to feelings of homeyness through a sense of freedom, mobility and comfort. Specifically, homeyness was enabled through an embodied sense of openness and flow tied to house size and design. It is in this intersection between flow and size that domestic practice becomes entwined in complex ways with the materiality of household sustainability.

Feelings of openness and flow were both a visual and whole-of-body experience. Visual flow was about the ways the space was encountered when entering or living in a space: it was about the ease through which people felt they could take in the space as the eye moves across its landscape of objects, furnishings, textures and light, and the feelings that this gave rise to, whether feelings of relaxation, comfort or anxiety. Feelings of visual flow were tied up with norms of domestic practice, particularly around dirt and tidiness. Flow could be restricted by clutter, described as an excessive number of decorative items, including photographs, as well as through the presence of overly detailed or 'fancy' objects, such as ornaments. Belinda maintained her own home according to a modern, minimalist aesthetic, but discussed feeling anxious when visiting houses that she felt were cluttered (see Figure 5.3):

> *I hate walking into a house and you feel like if you move you're going to knock something over because it's small and then they've crammed all these ornaments and things onto shelves everywhere and yeah, it's just like, it's like it's stressful, you don't know where to look because there's so many things ... it makes you start to feel tired in some ways.*

Participants also experienced their own homes this way, but expressed a more ambiguous relation to 'clutter', which typically consisted of significant family items like children's toys, photographs and ornaments. For example the clutter in Rose's current home was experienced as familial and homey, but also as making the space feel small. Rose and her husband were excited about moving to a larger house described as having a more 'sterile' look and feel: this home would absorb rather than simply contain the objects, transforming them from house-shrinking clutter and re-establishing their value as homey objects.

Flow and openness was also a whole-of-body experience, with feelings of homeyness associated with easy navigation and mobility through the house (cf. Imrie 2004 on disability and house design). Whereas visual flow entailed an encounter *with* objects, bodily flow took place in the spaces around and between the objects and walls inside the house. Feelings of flow were enabled through a sense of openness, as Jenny explained of her open-planned house: 'You don't feel

Figure 5.3   Spaciousness, Flow and Minimalism

boxed in, you don't feel confined to a certain room or certain area, you're sort of free to come and go.'

By contrast, flow was frustrated by smaller or less open spaces, particularly those that additionally performed as a thoroughfare. Flow thus took place around other inhabitants of the house, with children having a particular capacity to crowd and fill up a space. In June's previous house the kitchen was smaller *and* provided a route to the dining room; her children frequently filled up and ran through the kitchen while she was cooking. The new, more open-planned home sequestered the kitchen from family travel, providing a sense of sociality and openness where, when 'the children are here they can do their homework but in a separate bit, not on top of me when I'm trying to cook and run around the kitchen getting stuff out of the drawers and that'. The sense of openness supported familial togetherness while also affording June space to complete kitchen tasks. Rose likewise described her new 'nice big kitchen. One that you can actually walk in and walk out [of] so it's not cluttered, so you're not walking all over each other'. This sense of easy mobility was central to homeyness and provided an important motivation for moving house.

Senses of homeyness were also shaped by the flow of objects, sounds and light through the house, although these different elements often sat in tension with each other. Children and teenagers were a key challenge, with adults concerned to limit the movement of children's toys and noise throughout the house so that they could have child-free time and space or, alternatively, to minimise the spread of adult conversation into children's sleeping areas. For these purposes a large, closed-off house was valued, allowing toys and sounds, including conversations and television, to be contained in specific areas. Yet at the same time participants emphasised the importance of 'lightness', 'airiness' and warmth, observing that these are limited within a closed-off house. These conflicting ideals are clearly illustrated in Louise's comment:

> *I like to have light in a house, so I like the fact that there's big windows, the kitchen is fairly open-planned and I like that you can be in the kitchen and if people are out there in the family room you can watch, and you could still be playing with them or whatever even while you're cooking them tea ...*
>
> *[But] once we had kids we realised that in some ways it's a little bit too open. Because we've got the living areas all down one side and the bedrooms all down one side, like if we have people over, the bedrooms are very close and the kids can see and hear what's going on and sometimes you just want them to go to bed. [laughs]*

These everyday efforts to manage light, sound and family turned what initially appeared as an ideal house design into a challenging and fraught relation where the design placed restrictions on the ways in which the family could live in the house, threatening their sense of homeyness through an inability to effectively manage familial separation.

Senses of openness and flow, then, were central to residents' sense of homeyness. Yet these feelings were also mediated by relations with objects in the house (which readily transformed into clutter), as well as inter-familial relations. Strategies for managing these relations frequently referenced house size and design. Specifically, flow and openness were associated with largeness and expansiveness. Through contrasts between new and older homes participants emphasised the association between large housing and feelings of openness, freedom and comfort. Large houses enabled appropriate storage, which, as Belinda explained, allowed clutter and excess objects to be stowed out of sight. Storage meant: 'You can see what you have to get to' and 'it's easy to keep the house tidy because you've got places to put things.' In many cases storage exceeded cupboard space to also include 'spare' rooms and even the second storey of a house. Justine used a spare bedroom to house her and her husband's clothes, while Meg was one of a number who valued 'upstairs' for hiding washing, vacuum cleaners and other 'mess'. Large houses further facilitated the 'storage' of children, whose activities could be limited to specific sections of the house so that their toys and noises could be separated from the rest of the household. The purchase of large housing was a key strategy for managing openness and flow and resolving stagnation in many households; many of the participants had moved or were soon to move from smaller housing stock into these much larger homes which were seen to more readily accommodate the needs of a family.

A second strategy to achieve a sense of spaciousness and flow was through open-planned design complemented by careful colour choices and the strategic selection of housing materials, for example using glass instead of solid doors so that more light could enter the house. June explained:

> *It just feels more spacious. You don't feel like you're tripping over everything all the time. Whereas when you've got, see it might be the same amount ... of floor space as one that is open-planned, but you have to go through doorways and that to get there. So it's just nice that it's more ... it just feels more airy and just free.*

She especially emphasised the outcomes of such openness in terms of its impact on mobility. She demonstrated how, in a previous home that included a separate formal dining room, the physical necessity of squeezing past chairs when leaving the room, explaining that: 'It was nice to have the separate room but it didn't feel as spacious as it does now.' Or as Kathy captured:

> *We still wanted to create a sense of space and when you sit in the lounge room, there's no door so of course you look [towards the sitting room], and of course it creates an image of space. Whereas it's not, you know.*

Homeyness in these contexts, then, was associated with and achieved through a sense of spaciousness and the ability of objects and people to flow through the house.

And Household Sustainability?

While not directly engaging with debates around sustainability in this chapter, we have attempted to illustrate some of the ways in which people engage with and shape the objects in and materiality of their suburban, detached houses, and how this engagement is motivated by and shapes feelings of homeyness. In this short conclusion we summarise the connections between furniture, flows and homeyness, and sketch further connections to household sustainability.

In terms of furniture, we have suggested that keeping old items of furniture co-exists with purchasing new furniture, although importantly such practices are framed by both aesthetic and familial considerations. It was deemed important that houses felt 'lived in' rather than empty and sterile, and this could be achieved if they were filled with objects that had familial connections or had been acquired through friends and acquaintances. Crucially, these accounts point to fissures in the popular scripting of McMansions as the epitome of unsustainable practice. In the case of furniture this fissure can be detected in the economy of second-hand furniture that is central to everyday familial relations and practices of home. These furnishings are not only connected to feelings of home, family and belonging but also, in terms of sustainability, offer an alternative to new purchases. Thus it is the case that eco-efficient/sustainable practices co-exist alongside more resource intensive ones. Furniture provision associated with creating a homey space drew upon both re-circulations and new purchases. This suggests that broader, sustainability-oriented concerns with reuse may have traction in these suburban environments if aligned with everyday practice and constructions of homeyness.

In terms of openness of flow, we have suggested that participants achieved these goals for home in two ways. The first was dependent on physical largeness, materially idealised in the form of the two-storey home with multiple, open-planned living spaces. The second involved a sense of spaciousness and flow through better design. In the mass-designed project homes visited in this research the former has a more clearly antagonistic relation to ideals of domestic sustainability and eco-efficiency: these large houses were typically dependent upon artificial air-conditioning and heating to achieve comfort; some were designed without eaves to protect the house from the sun; and one, built before changes in government legislation, did not have any insulation. This type of house is a significant component of new domestic supply in Australia and was selected by those who participated in this research as a solution to issues of homemaking, rather than out of a concern for environmental outcomes. Openness and flow gave rise to feelings of homeyness that participants were less likely to have in more cramped and cluttered spaces. It is in the second option that opportunities for change are more clearly evident. In creating a 'sense of space' through better design and the careful selection of building materials that enable air and light to flow into the house, these houses are suggestive of more eco-efficient design: of passive heating, cooling and lighting coupled with a smaller physical footprint, while still affording the sense of spaciousness that has become so central to the ways in which people imagine and live home and family, and thus choose their housing. Although there is no simple relation between smaller housing and eco-efficiency, the domestic practices of people who took part in this research suggest that aligning the dual imperatives of homeyness and sustainability in the materiality of housing might be a key initial step in the move towards more sustainable household practices.

References

Askew, L.E. and McGuirk, P.M. 2004. Watering the suburbs: distinction, conformity and the suburban garden. *Australian Geographer*, 35(1), 17–37.

Blunt, A. and Dowling, R. 2006. *Home*. London: Routledge.

Clarke, A.J. 2002. Taste wars and design dilemmas: aesthetic practice in the home, in *Contemporary Art and the Home*, edited by C. Painter. Oxford and New York: Berg, 131–151.

Curtin, J. 2009. How the McMansion super-sized the suburbs. *The Sydney Morning Herald*, 5 December, 7.

Dowling, R. 2008. Accommodating open-plan: children, clutter and containment in suburban houses in Sydney, Australia. *Environment and Planning A*, 40(3), 536–549.

Gorman-Murray, A. 2006. Gay and lesbian couples at home: identity work in domestic space. *Home Cultures*, 3(2), 145–168.

Gregson, N. 2007. *Living with Things: Ridding, Accommodation, Dwelling.* Wantage: Sean Kingston Publishing.

Hetherington, K. 2003. Spatial textures: place, touch, and praesentia. *Environment and Planning A*, 35(11), 1933–1944.

Hitchings, R. 2004. At home with someone nonhuman. *Home Cultures*, 1(2), 169–186.

Hobson, K. 2006. Bins, bulbs, and shower timers: on the 'techno-ethics' of sustainable living. *Ethics, Place and Environment*, 9(3), 317–336.

Imrie, R. 2004. Disability, embodiment and the meaning of the home. *Housing Studies*, 19(5), 745–764.

Kaika, M. 2004. Interrogating the geographies of the familiar: domesticating nature and constructing the autonomy of the modern home. *International Journal of Urban and Regional Research*, 28(2), 265–286.

McCracken, G. 1989. 'Homeyness': a cultural account of one constellation of consumer goods and meanings, in *Interpretive Consumer Research*, edited by E. Hirschman. Provo, Utah: Association for Consumer Research, 168–183.

Miller, D. 2002. Accommodating, in *Contemporary Art and the Home*, edited by C. Painter. Oxford and New York: Berg, 115–130.

Moran, S. 2008. Under the lawn: engaging the water cycle. *Ethics, Place and Environment*, 11(2), 129–145.

Noble, G. 2004.) Accumulating being. *International Journal of Cultural Studies*, 7(2), 233–256.

Power, E.R. 2009a. Domestic temporalities: nature times in the house-as-home. *Geoforum*, 40(6), 1024–1032.

Power, E.R. 2009b. Border-processes and homemaking: encounters with possums in suburban Australian homes. *Cultural Geographies*, 16(1), 29–54.

Rose, G. 2003. Family photographs and domestic spacings: a case study. *Transactions of the Institute of British Geographers*, 1(28), 5–18.

Shove, E. 2003. *Comfort, Cleanliness and Convenience: The Social Organization of Normality.* Oxford and New York: Berg.

Sofoulis, Z. 2005. Big water, everyday water: a sociotechnical perspective. *Continuum: Journal of Media and Cultural Studies*, 19(4), 445–463.

Chapter 6

Remaking Home: The Reuse of Goods and Materials in Australian Households

Ralph Horne, Cecily Maller and Ruth Lane

Introduction

This chapter explores the reuse or recycling of goods and materials within three domains of practice: food and beverage provision, home furnishing, and home improving and maintenance. Households in developed countries such as Australia experience significant flows of materials and resources, some on a daily basis and some less frequently. However studies of household consumption in environmental sustainability research have tended to focus on purchase decisions without exploring the ways in which goods and materials are used in the home or how they are eventually disposed of. Exceptions include Daniel Miller's work on household consumption (for example Miller 2001), Gay Hawkins' work on the ethics of waste (Hawkins 2006), Kersty Hobson's work on ethics of consumption (Hobson 2006) and Elizabeth Shove's work on practices involving consumption (for example Shove and Southerton 2000). Notwithstanding these examples, material flows within households and their environmental implications remain poorly understood.

In recent decades increased public debate around peak oil and climate change has coincided with increasing environmental concern among Western households (Australian Bureau of Statistics 2008, European Commission 2008) and policy shifts in waste management towards landfill diversion. Households are encouraged to take responsibility for environmental issues by changing their consumption habits and behaviour in the home. This approach, combined with neo-liberal ideas of market mechanisms, has formed the basis of many proposed policy solutions and is referred to by Hobson (2002: 97) as the 'rationalisation discourse of sustainable consumption'. Similarly, 'ecological modernisation' posits that sustainable consumption is to be achieved through consumers choosing to purchase more energy-efficient or 'environmentally friendly' goods (Darier 1996, Shove 2003, and see the chapters by Hobson and Scerri in this volume).

The increasing emphasis on household responsibility for environmental goods has occurred in conjunction with increased privatisation of previously state-owned utilities and resource management services. The move towards privatisation and the dissolution of monopolised utility markets implies households are being given a more specific role in the environmental management of these networked systems (Chappells et al. 2000). However what this role entails has been overlooked and underestimated in both scientific and policy fields (Chappells et al. 2000). In contrast, other socio-ecological systems where private and public/government system components intersect, such as agriculture or natural resource management (NRM), have been the subject of a long history of research and policy making (Pretty and Ward 2001, Carlsson and Berkes 2005).

Our concern here is what happens inside the home. We seek to develop an understanding of goods and materials without assuming that householders are rational or independent actors, but rather, by observing how households use goods and materials in carrying out daily practices and homemaking projects. Householders are increasingly being encouraged to make changes to food and beverage consumption, transport, energy and water use, as well as home maintenance and improvement (for example through measures such as retro-fitting insulation). However questions remain as to how environmentally sustainable outcomes might be reliably predicted from planned adjustments to existing arrangements. This trend raises a range of issues concerning the current and potential role of materials recycling and reuse in households as a response to both environmental imperatives and increasing scarcity.

We acknowledge a range of disciplines, including cultural studies, sociology and human geography, that have contributed theoretically to understanding the phenomenon of household consumption and the material culture of the home. Included among the various theoretical approaches are modern and post-modern concepts of lifestyle (Gram-Hanssen and Bech-Danielsen 2004); phenomenology; theories of culture and social action (Goodsell 2008); structuration theory regarding the interdependency of technologies, infrastructure, rules, and institutional and power arrangements (Giddens 1984); and social practice theory, which has as its basic premise a focus on the 'practical logic' of ordinary life (Bourdieu 1990, Hand, Shove and Southerton 2007, Gregson, Metcalfe and Crewe 2009). Collectively these approaches, and theories of practice in particular, have moved significantly beyond culturalist perspectives – which simply sought to ascribe meaning to objects and actions – to explore the dynamic and shifting relationships between people and things, and 'having and doing' in routines of everyday life (Shove et al. 2007). With a focus on the practical logic of ordinary life, theories of practice have considerable application in our attempt to explore recycling, use and reuse of goods and materials in households.

Theories of practice vary. For key overviews see Daniel Miller (1998), Andreas Reckwitz (2002a) or Alan Warde (2005). It is not our intention to provide a robust account of the different varieties of practice theory; rather we seek to briefly review common aspects of practice theory that can be usefully employed in empirical work. The constituents of a practice and how they are implicated in practice performance also vary according to different theorists. What does and does not constitute a practice is left open for interpretation by empiricists. However there is some consensus around the idea of practice as routinised activity which involves connected elements of or nexuses between bodily and mental activities, objects/materials and shared competencies (knowledge, skills). Further, various practices are interconnected with each other (for example practices of cooking linked with practices of shopping for food) and occur within wider political, economic, legal and cultural contexts of varying formality (Røpke 2009). Andreas Reckwitz (2002b) finds that the mental activities of understanding, skills and desire are qualities of a practice, not of the individual who performs them, and that conventions and standards of practice appear to steer human action rather than the other way around.

Although studies have explored social practices specifically in relation to the consumption of goods and materials (see for example Miller 2001, Shove and Hand 2005, Hand, Shove and Southerton 2007, Shove et al. 2007), none have examined the phenomenon of flows of second-hand goods and materials through households. In their article on home extensions in the United Kingdom, Martin Hand and colleagues focus on the accumulation of consumer goods and the social, temporal and spatial fabric of daily life (Hand, Shove and Southerton 2007). They argue that the relational dynamics of aspirations, practices, objects and materials is generally underemphasised: 'Little attention is paid to the way in which changing expectations and visions of domestic practices relate to the accumulation of material goods, and how together they (re)configure domestic spaces' (Hand, Shove and Southerton 2007: 670). Nicky Gregson and colleagues (2007a) welcome these contributions but note that they still tend to interpret consumption as accumulation and argue that understandings of consumption need to also include practices of circulation and divestment. In shedding light on the relations between practices and consumption, the following points are made: firstly, competencies (skills and knowledge) play an essential role in the performance of practices and how goods and materials (and technologies) are treated, and the level of knowledge and skill required varies depending on the practice (Hand, Shove and Southerton 2007); secondly, practices are not static, nor do they occur in isolation, and can be 'structured' by past experiences and future visions (Shove et al. 2007); thirdly, changing practices associated with accumulation, circulation and disposal of material goods can reconfigure domestic spaces (Hand, Shove and Southerton 2007).

In this chapter we focus on practices and activities involving the use of goods and materials in residential spaces, and the associated competencies, spatial and temporal dimensions and systems of provision associated with the flows of these materials through the household. Drawing on the ways in which both routine and technological structures inform consumption (Shove et al. 2008, Watson and Shove 2008), we specifically direct the focus on changing practices around the substitution of 'new' goods and materials with second-hand alternatives. In our conclusion, we consider the potential for the incorporation of used goods and materials in households within emerging governance arrangements and systems of provision to eventually achieve a transition to more materially sustainable patterns of living.

Approach and Method

It is an ontological assumption of this work that the substitution of second-hand goods and materials in households for new goods, and reuse of goods and materials purchased new, that is, 'product life extension' (Cooper 2005), is a desirable environmental goal for reducing waste and other environmental impacts associated with consumption. In our three selected domains of practice of food and beverage provision, home furnishing, and home improving and maintenance, various opportunities are encountered for the reuse of goods and materials, the substitution of new goods and materials with second-hand alternatives, or divestment and disposal. For example in the first domain, food packaging that is no longer required can be recycled, reused or disposed of.

Each of the three domains is associated with particular sets of practices: the first domain is associated with practices of cooking, eating and perhaps socialising, while the latter domains are associated with other homemaking practices. Within the broader realm of practices, our focus is on the particular points of intersection between actors, objects and materials evidenced through 'doings and sayings' (Schatzki 2002). In taking this approach we make the assumption that practices are central to consumption and that changes in practice present specific opportunities for the reuse or substitution of goods and materials. In this chapter we address two questions:

1. As householders undertake food provision, furnishing and home improvement and maintenance, what skills, spatial and temporal characteristics, and systems of provision are involved when second-hand goods and materials are substituted for new?
2. What factors assist or hinder the reuse or recycling of goods and materials at the household scale and how can this knowledge inform new policy approaches aimed at facilitating more widespread reuse practices?

We draw on interviews with householders undertaken with three cohorts of participants. The cohorts were recruited for three separate but related studies, but because in all three we observed householders making use of second-hand goods or materials. In this chapter we have 'scaled up' the data (Valentine 2006) by combining and assessing the three data sets together.

Householders in Cohort 1 were home-owning Australians who had undertaken some form of home improvement in the last year (renovation, retro-fitting or both) and were planning to initiate future home improvement projects. The 16 householders from the Australian states of Victoria, South Australia, New South Wales, Western Australia, Queensland and Tasmania were randomly recruited by a professional recruitment agency and interviewed between October and December 2008. The purpose of interviewing homeowners in this first group was to understand the context of their home improvement and maintenance projects, their attitudes towards and concerns for the environment, and activities undertaken within the home, including recycling. Householders in Cohort 2 provided a cross-section of household configurations, housing types and tenures, with households being pre-selected as conversant with the use, acquisition and disposal of second-hand goods. Six householders in Victoria were interviewed in their homes in early 2009 in conjunction with observations of home configurations and practices associated with recycling and reuse. Cohort 3 consisted of 16 self-identified 'green' renovators in Victoria who had undertaken a renovation in the last two years. One or two representatives of each household were interviewed in their dwelling and photographs taken of the house and garden during walk-through tours, conducted between April 2008 and March 2009.

Across the three sets of data, householders were well-educated, with most possessing a tertiary degree and many possessing postgraduate qualifications. Incomes varied, although most households contained at least one person in full-time employment. Professions ranged from public service, health and community work to corporate and private consultancy. The age and style of the houses occupied varied greatly and included post-World War II weatherboards, an ex-1956 Olympic Games village concrete home, brick 'Californian bungalows', and townhouses less than ten years old, but in general most householders occupied older, detached houses with a garden.

In our data we explored activities involving materials and goods within the chosen household domains across the three cohort data sets, focusing on the following themes:

- Competencies (skills, knowledge and planning)
- Interactions with informal and formal systems of provision
- Spatial and temporal factors.

For this purpose, 'competencies' cover the skills, knowledge and preparation required for the acquisition, treatment, use or reuse, disposal and divestment of goods and materials in the home. Planning, although generally considered a type of skill, is mentioned specifically because of the integral role it plays in activities involving reuse of goods and materials. 'Formal systems of provision' include recognised systems associated with markets and exchange of money or government services and infrastructure. To some extent formality also has a scalar dimension, with 'informal' referring to the household scale or the scale at which interpersonal networks exist, and 'formal' being a more anonymous interaction between retailers and buyers or service providers and clients. 'Spatial and temporal factors' are the space and time requirements involved in the acquisition, treatment, use or reuse, disposal and divestment of household goods and materials.

Our framework for analysis was bounded by the domains and themes related to particular goods or materials. While the three data sets were initially analysed separately, we then combined the results to explore shared themes across the various household domains. In our first domain of food and beverage provision we focused on the recycling of packaging and associated disposal or composting of unwanted food. In the second domain, home furnishing, we explored the acquisition, use and divestment of used furniture. In the third domain of home improving and maintenance we examined the acquisition and use of recycled construction or building materials and appliances. In the following discussion the three data sets are combined where applicable and the analysis reported by domain.

Food and Beverage Provision: Packaging and Food Composting

Within the domain of food and beverage provision, packaging is used to store and transport food and beverages and, invariably, households accumulate packaging waste. Over recent decades a range of different systems of provision have been organised under the general framing of 'domestic waste management' in an effort to divert fractions of this waste from landfill. Over the same period, domestic practices of waste management have changed in response to both the socio-cultural contexts in which household food and beverage provision takes place and the waste management systems provided.

Competencies (Skills, Knowledge and Planning)

When the householders in Cohort 1 were asked to describe efforts to reduce consumption, recycling was the most commonly reported action, with 12 out of 16 householders reporting that they recycled paper and plastic packaging. Consistent with national data on recycling (Australian Bureau of Statistics 2006), this appeared to be the most frequent and consistent effort households made which they identified as contributing towards reducing their consumption or its environmental impacts. Where it was not mentioned, it was probably because it has become an ordinary part of daily life and not necessarily considered worthy of mention. Householders discussed other steps they were taking to manage food waste at home, such as diversion of organic food matter to composting. Practices such as recycling and composting are reinforced through formal education channels, where children are encouraged to play an active role in household recycling practices:

> *You know, I recycle and I've taught my children to recycle. We don't waste. It's easier now, it's in the schools. You know, my eight year old tells me to put something in the recycling bin. Yeah, it's good.* (Cohort 1, Householder 15)

Practised recyclers deployed very particular skills and judgements to determine what constitutes a competent recycling activity. Some of our householders clearly had a conceptual framework that incorporated food selection according to packaging – in other words, the amount of packaging implied at the input end, as well as the practice of dealing with it in the kitchen:

> *I'll buy six apples instead of a bag of apples, I'm conscious like that. When I do purchase things I would open up the packet and look at what could be recycled,*

so I will squash the cardboard and I'll put it into the recycled paper and I'll put the plastics into a recycle thingo so maybe a tiny bit of cellophane goes into the bin. (Cohort 2, Householder 1)

This might be accompanied by an approach to shopping that entails carrying cloth shopping bags and favouring stores where loose items can be purchased.

Interactions with informal and formal systems of provision

Household efforts around recycling containers also appeared to be influenced by the infrastructure provided by government agencies. For example South Australia has container deposit legislation and associated collection systems, and one South Australian householder discussed how she regularly delivered her tin cans and glass bottles to a recycling depot where they were paid for each item:

Householder 5: *And cans and bottles, I keep them now and you take them up to the depot.*
Interviewer: Is it far away the depot?
Householder 5: *No, Angaston.*
Interviewer: How often would you do that?
Householder 5: *Only when the bin gets full, I took it the other week and it was the best day, it [the price] went up to 10 cents.* (Cohort 1)

In this case, the formal system provided for managing containers required skills and resources in terms of sourcing drop-off depots, planning and making trips and organising transport. These requirements were less demanding for the majority of householders across the cohorts who were based in Victoria. Here the formal system of provision involved co-mingled recycling bins and their regular collection by local councils, so that most householders needed only to collect their recycling materials in the process of preparing meals and deposit them in their bin, which is then placed on the kerb for regular collection. As the kerbside schemes used in Victoria place fewer requirements upon householders than so-called bring schemes, such as the system in South Australia, we might expect higher rates of recycling. However the rates of recycling participation reported by households do not differ greatly between Victoria and South Australia (Australian Bureau of Statistics 2006). We attribute this to our observation that in both states a diverse range of informal recycling systems interweave with these formal systems and therefore affect the rate of reported participation in recycling activity. This includes dedicated infrastructure arrangements within the home (such as interim recycling bins for specific materials), negotiation, organisation and division of tasks between householders, and specific activities associated with sorting and preparing packaging for recycling associated with food and beverage provision.

Spatial and Temporal Factors

Several households maintained significant informal infrastructure to support sophisticated arrangements which they themselves tended to refer to as their recycling 'system' – effective means of sorting their packaging and food waste:

> *We have a great composting system so all our food scraps are recycled. I collect paper from work and that goes straight into the compost bin. We use the standard recycling process. We have our own green recycling process.* (Cohort 1, Householder 3)

Apart from skills and planning, such arrangements involve temporal and spatial considerations. Time was not mentioned as a significant constraint in connection with food and beverage container recycling or composting, suggesting that the (sometimes considerable) time invested by some households is successfully incorporated into daily patterns of living. Similarly, while the space used for such systems was sometimes considerable, this was only the case where space was available, such as in a large kitchen or a detached dwelling with a sizeable backyard, side areas, garaging or frontage.

Thus in food and beverage container recycling, formal systems of kerbside bin collections for recycling food packaging waste have co-evolved with practices of packaging recycling in the general population which are associated with basic skills, knowledge and competencies at the household scale. Furthermore, the formal system appears to have inspired, or has been extended by, some households to create their own informal household systems, for example, to recycle food waste. Householders' reference to particular home recycling or composting practices as part of 'systems', and the way these household-scale systems were differentiated from formal infrastructures for waste management, indicates that householders employed a considered approach involving organisation and commitment.

Home Furnishing: The Acquisition, Use and Divestment of Used Furniture

Unlike food, home furnishings are a more enduring aspect of the materiality of everyday life. Items of furniture and decor are occasional acquisitions rather than regular ones, and divestment of such goods is also spasmodic. Second-hand acquisitions may be spontaneous and opportunistic or carefully considered, and combinations of second-hand acquisitions alongside new purchases are common. Items of furniture and other household goods may be acquired through informal means such as scavenging materials discarded by others, through inheritance, as gifts or exchanges between family and friends, or through any of a range of retail outlets ranging from online retail sites such as eBay™ to inexpensive charity shops to upmarket antique stores (Lucas 2002, Gregson and Crewe 2003, Gregson 2007, Gregson, Metcalfe and Crewe 2007a, Gregson, Metcalfe and Crewe 2007b). As with other forms of shopping, the acquisition of second-hand furnishings can be a complex exercise that connects with personal emotional states, with expressions of identity or social status, or with expressions of care for significant others (Miller 1987, Miller 1998). Given that we already know that such shopping is embedded in social relationships among families and friends, we would expect those incorporating second-hand goods into their home furnishings to view this as acceptable within their social milieu.

Competencies (Skills, Knowledge and Planning)

Interviewees across Cohorts 2 and 3 generally displayed a sophisticated level of resourcefulness and initiative in learning the skills needed to acquire, repair or dispose of furnishings. For example one of the renovator couples from Cohort 3, who were semi-retired, learnt how to use the Internet to dispose of existing kitchen and laundry fittings and became competent practitioners in this regard. Another family in Cohort 2 used the *Trading Post* to sell their old kitchen, removed when renovating, to a family who planned to install it in a holiday house.

The second cohort comprised a cross-section of households at different life stages and with different housing tenure arrangements and, as such, provides some indication of how second-hand furnishing-related skills, knowledge and planning may vary over the life course. Householder 1 in Cohort 2 was a recently graduated student sharing a house with friends who had enterprisingly acquired several pieces of second-hand furniture for free. These had been put into storage and attributed with significant sentimental value.

While some people clearly treasure objects for their impression of a past history of ownership, Householder 2 in Cohort 2, who came to Australia from India five years before our visits, acquired second-hand furnishings at least partly out of economic necessity. When he and his wife arrived in Melbourne as students they had no furniture at all and sourced most of the furnishings for their rental flat second hand, exercising careful financial planning. For this couple, the use of second-hand furniture was associated with a life stage in which their main objective was to save money. However Householder 2 expressed a strong aversion towards the idea of scavenging used furnishings from hard rubbish collections:

> *I would rather not have a chair in my house than to pick it up from the street. I would go out and make money and then buy it because it would make me feel heaps better if I know that ... this chair gives me a positive feeling and [a chair reclaimed from hard rubbish] would make me feel worse.* (Cohort 2, Householder 2)

This aversion did not apply to receiving a second-hand item as a gift from a friend. However ultimately he hoped that in the future they would be able to buy most furnishings new. They anticipated having a child and buying a house, after which a return to purchasing second-hand goods would feel like slipping backwards. However he hoped to maintain his current anti-materialist ethic of only acquiring what was needed and avoiding waste and clutter, and related this to environmental concerns.

Householder 5 and her husband in Cohort 2 had purchased their house eight years earlier and had recently undertaken an extension to better accommodate the work space and storage needs of two working adults and four school-aged children. The whole family were active in acquiring and disposing of second-hand goods and some members took an interest in repair and maintenance using sewing or carpentry skills and equipment. Through relationships developed with the proprietor of a local second-hand goods store this householder swapped a good quality dresser for a second-hand dining table. Her family had also acquired items of discarded furniture from the nature strip in their local streets and actively sought out reuse channels for disposing of items they no longer needed.

Skills and knowledge for repair and maintenance work were perceived by some to be more closely connected with older generations. Householder 3 in Cohort 2 felt that the disposition she and her husband had towards repairing and extending the lifespan of household furnishings was more strongly associated with their own generation (60 years plus) than that of their children or grandchildren. They had been in their current house for 35 years and had acquired furnishings through a mix of new and second-hand purchases. While they clearly had an eye for value for money, some second-hand purchases were made to fit with the style of existing furniture. Her husband worked part time for a real estate agent clearing out houses so that they could be rented out. He often brought home small items of furniture that he could not bring himself to take to the rubbish tip. These were usually given away to friends, community groups or schools. A recently widowed retiree in Cohort 2 had never acquired household furnishings second hand but consciously prolonged the lifespan of all his possessions using his carpentry skills and equipment. He had recently given some of his carpentry tools to one of his sons, who had trained in cabinet making. However the son had not made much use of them due to the demands of long working hours and children.

Interactions with Informal and Formal Systems of Provision

Channels for divestment of used furniture varied in their degree of formality and included the use of annual council hard waste collections, donations to charity organisations or gifting to family or friends. While the recycling or reuse of food waste and packaging required a level of organisation and 'systems' thinking, the reuse of home furnishings appeared to require a knowledge of suitable retail outlets in addition to the expected practical mending skills, tools and convenient spaces for storage and restoration work. Across our cohorts, those who were active (or had been) in acquiring and using second-hand furniture were generally adept at identifying appropriate formal and informal channels. However, this should not be regarded as evidence that second-hand furniture channels are easy or convenient to learn and navigate.

Also, some householders had made use of formal waste management services and facilities (for example hard rubbish collections, tip shops and so on) to acquire or dispose of furniture but none were solely reliant on government waste management services, with networks of 'friends and family' playing a significant role.

Spatial and Temporal Factors

Having had a very mobile life as a student, moving between Australia and Malaysia, Householder 1 in Cohort 2 had only recently begun to acquire things as she anticipated renting her own flat in the future. However, much of her furniture was stored with family members as her housemate preferred a more modern style. This highlights the fact that furniture is often 'shared' and negotiated amongst household occupants with different needs and wants. The reasons for long-term storage of furniture appear to extend beyond notions of economic value or sentiment. In this case it was not just the fact that it was this householder's first couch being stored; it also signified the accumulation of home furnishings as part of the journey towards a vision of the future which involved a privately owned dwelling with her own possessions. However there may also have been a temporal inertia at play; once stored, the perceived time and effort required to 'un-store' and divest could be a deterrent. Either way, even as an itinerant young migrant, Householder 1 had overcome lack of space and sought to acquire and retain her second-hand furniture.

In our second cohort the presence of children generated additional demands for space and quantity of household furnishings and affected how people related to household goods and materials more generally, a point also made by Gregson and colleagues (2007a) in relation to UK-based research on divestment as an aspect of consumption. One young couple on a modest single income with three children under five explained that since becoming parents their relationship with goods had shifted from an 'ownership' to a 'utility' focus and they had become quite comfortable with sourcing clothes, furniture and other goods for both themselves and their children through second-hand channels. This was also influenced by frequent house moves and living in rental properties.

Lack of time sometimes appeared to affect the capacity of younger generations to undertake repair and maintenance work. Others placed particular value in avoiding new purchases of furniture and were willing to invest considerable time and space to this end. This seemed odd in the case of one householder who had embarked on an expensive extension incorporating a living and dining area and a new kitchen but expressed considerable pride in the fact that she did not buy any new furniture:

> *Another thing I was really happy about though, if I can just tell you; I didn't have to buy any furniture; all the furniture is old furniture of the family or stuff that I have stored or bought second-hand or hard garbage or something like that.*
> (Cohort 3, Householder 3)

In this case access to storage facilities was clearly an important factor in enabling this level of reuse.

This brief cross-section provides some indications of how and why the reuse of second-hand furnishings might vary over the course of one's life. In this respect our research confirms findings of similar research in the United Kingdom (Gregson, Metcalfe and Crewe 2007a, Alexander et al. 2009a, Alexander et al. 2009b). Younger people often have a more transient lifestyle, which makes it difficult to own and store furnishings. They are also likely to be driven by economic considerations and seek out second-hand furnishings because they are cheaper. Households with children may turn over furnishings more rapidly as the needs of their children change. Retirees or those nearing retirement age have less need to acquire furnishings and are more likely to be divesting themselves of items they no longer need.

Few of our interviewees suggested that lack of tools or skills, or lack of knowledge of where to acquire or dispose of second-hand furniture, was a constraint on their reuse of furniture, and they seemed adept at acquiring new skills if they felt they needed them. They were more likely to emphasise lack of time. However limited access to storage space also featured in some accounts.

Home Improving and Maintenance: Recycled Building Materials and Reused Appliances

Daily home maintenance practices include cleaning, arranging and organising internal spaces to enable the range of home services to be achieved. However notions and expectations of what constitutes an acceptable internal space and décor are diverse and continually shifting (Attfield 2000). We recognise that in many households, competence as a homemaker is widely regarded as requiring *both* regular attention to cleaning and maintenance *and* more episodic upgrading of equipment and spaces to meet changing needs and visions of how things should be (Shove and Hand 2005). Where once homes and their interiors were built and designed to last a lifetime, there are now increasing expectations, aspirations and opportunities to modify homes on a more frequent basis, such that a once episodic activity is now regularly and habitually reproduced and is widely embedded within home-owners' cultural practice. Attendant technologies and practices co-evolve with the result that new demands are made on homes and the objects and spaces within them, which in turn influence our experience of space and the value placed on different physical configurations (Hand, Shove and Southerton 2007). We may expect that in some cases changed homemaking expectations are satisfied within existing spaces while in others homemakers conclude that they simply need more room or that improvements are required which involve significantly reconfiguring, upgrading or otherwise renovating the existing internal space.

Rather than focusing on daily maintenance, in this stage of the analysis we explore the more episodic activities we term 'home improvements', in the process examining the role of second-hand goods and materials. Hence our attention turns to understanding how households accommodate shifting visions and expectations of homemaking within domestic spaces. Therefore, rather than seeing kitchen reconfiguration as part of the domain of food provisioning practice, we deliberately place it within the realm of homemaking, understanding that kitchen reconfigurations relate to the realisation of 'visions and expectations' rather than food preparation per se (Hand, Shove and Southerton 2007).

Home improvements are means to enhance a dwelling, resulting in benefits such as increased comfort, health and well-being, status and market value. Many owners buy a house with the intention of making changes (Baum and Hassan 1999, Allon 2008) and home improvements are a large-scale phenomenon which may typically 'require' an increase in dwelling size or number or size of appliances (washing machines, televisions, microwave ovens, air-conditioners and so on). In their work on kitchen renovations, Elizabeth Shove and Martin Hand propose that renovations and kitchen-based practice are structured by past experiences and by images of the future: '[I]nvestments in new appliances and in kitchen make-overs were commonly desired, anticipated or justified as a means of bridging between dissatisfactions of the present and an image of a better, more appropriate future' (Shove and Hand 2005: 4).

Depending on factors such as the nature and extent of the work, and the skills and knowledge required, home improvement can be carried out in two main ways: by engaging building professionals (including architects, designers, builders and tradespeople) or 'do-it-yourself' (DIY), with the acknowledgement that 'hybrids' of the two are also possible. In Cohort 1 the majority of home improvement projects homeowners undertook were retrofits rather than renovations, while in Cohort 3 nearly all homeowners extended their homes or added second storeys.

Competencies (Skills, Knowledge and Planning)

There were few opportunities in the retrofits undertaken in Cohort 1 for householders to consider using recycled materials or goods; hence our analysis is limited in this respect. There were two exceptions. One householder had sourced a second-hand stove for her kitchen, commenting on her own skills in thriftiness and the wastefulness of those who only source new products and appliances:

> *I actually got a second-hand stove and top. Because that's another thing, I find how people re-do their kitchens so often [alarming]. Oh my God! The money that they spend, just to be up to date. I think it's quite decadent. And the stove I [bought], it was hardly used. And I got it for $60, you know. And my gas top didn't have a mark on it ... someone had just taken it to the tip.* (Cohort 1, Householder 10)

Another householder indicated her knowledge and competence in understanding energy flows associated with materials, explaining how she intended to source some second-hand or recycled-content window frames, acknowledging the high amount of embodied energy in some materials:

> *The windows that I'm going to put in, not all of the windows, but a couple of the windows, are going to be recycled and the ones that aren't going to be recycled I'm considering the sort of materials that are being used for them. So I don't want to get aluminium windows or at least not fully aluminium windows. I'd like to get windows that have a better thermal conductivity and perhaps have less of an energy embodied cost to them.* (Cohort 1, Householder 13)

In reflecting on how the recycled timber finish was achieved in a living room, one interviewee commented that the skills and competencies required in sourcing, preparing and installing recycled timber were akin to those required in a trade:

> *My father was a tool-maker so it's sort of like doing work properly and carefully.*
> *Tim will have a go at anything that breaks in the house because he's sort of ... self-taught [having the skills to be able to do a] bit of carpentry, [he's a] bit of everything because he's got a brother in the trade.* (Cohort 2, Householder 2)

The interviewees' accounts of their involvement in home improvements suggest that to successfully incorporate recycled materials in home maintenance and improvement, regardless of professional or DIY home improvement projects, substantial competencies are required, along with planning and the ability to navigate both formal and informal systems. Similarly, in a study of four families undertaking DIY home improvements, Todd Goodsell (2008) found that as they worked through their improvements and engaged with materials of varying kinds, householders' skills and knowledge were developed 'on the job', implying that, initially, competencies to undertake the work were low. Largely through reliance on informal networks of neighbours and family, their knowledge and skills grew to the point where one householder became an independent contractor specialising in home restoration (Goodsell 2008).

Interactions with Informal and Formal Systems of Provision

Four out of the 16 households in Cohort 3 who said they were doing a 'green' renovation did not use recycled or second-hand materials. These householders instead focused on sustainable technologies (such as grey water recycling and solar hot water systems) rather than materials used in the building fabric. The remaining householders in this cohort were primarily concerned with the use of recycled materials as well as the disposal of unwanted materials, and for some, household goods (for example recycling of kitchen cupboards, concrete and steel). They actively engaged with formal systems of collection and disposal, sometimes going to considerable time and effort.

However even those who thought they were competent and had the time encountered difficulties attempting to dispose of some materials, for example:

> *Oh I spent three hours on the phone one afternoon ...I even rang a few people in the local rag to say 'Look you pick up and remove rubbish, what do you do with paint?' 'Oh it just goes to landfill.' I said, 'No sorry, that's not legal' and he said 'Oh well, that's what I do', and then hung up [laughs]. So it is tricky ... some of the systems are [still] not fully in place and I was somebody who knew the processes to find to get rid of these things, but even I had difficulty.* (Cohort 3, Householder 5)

The most popular recycled material was timber which was used in flooring, doors, windows and cupboards, pelmets and, in one case, as a source of fuel. Other materials from the building structure, such as bricks and plaster, were reused in the reconfiguration of walls and interiors. Some householders also sought products with recycled content, such as concrete for building slabs, insulation and decking. Although householders often intended to use recycled materials and had the skills, competencies and planning to source them, the patchy nature of the system underlying the building industry resulted in their plans sometimes being thwarted:

> *I found a place with the recycled timber and someone had recommended a builder to us who could do the job straight away and was cheap. And he just said 'I really really really really really really really don't want to lay recycled timber'. He said, 'It takes way longer. You're going to have to pay labour costs, it's not tongue and grove at the ends, only at the sides. I have to chop it off every time I get to a joist and add another bit...' In the end...the timber I got from the same place ...that sold the salvaged timber but it was supposedly from a sustainable plantation.* (Cohort 3, Householder 16)

Spatial and Temporal Factors

This problem also seemed to occur when households did not have enough time to research products and materials:

> *It's very tricky ...when you're dealing with builders and you have to make decisions on the run ... when you've got five blokes sort of looking at you saying, 'But we need the timber for this next [job] and we've moved on to this, and we need this next week'.* (Cohort 3, Householder 6)

However, when they succeeded in negotiating these systems, householders were proud to point out the effort they went to, to obtain their recycled materials.

> *This is all recycled from all over Melbourne ... So look we had a team, one team*
> *putting it up and another team preparing, and [my husband] and I getting it from*
> *all over Melbourne, looking up the* Trading Post *and things, spending Saturday*
> *mornings every week. So look, I'm a wood freak ... we didn't realise how hard*
> *that would be, did we? Or how expensive it would be, but we're delighted we*
> *hung on to the concept, aren't we?* (Cohort 3, Householder 14)

The majority of retrofits in Cohort 1 did not require such extensive competencies, were relatively less time intensive, and did not involve complicated formal or informal systems of provision. It was the renovations undertaken in Cohort 3 which required high-level competencies and planning, were time and space intensive, and were highly dependent on both formal and informal systems. Not all householders in Cohort 3 paid professionals to do their home improvement work. Several actively developed high-level competencies and were willing to accommodate the spatial and temporal requirements by carrying out the majority of the improvements DIY. These competent practitioners also successfully negotiated the formal and informal systems required to achieve their 'green' renovation, often taking difficult routes to obtain or reuse building materials.

The level of skills and knowledge involved in undertaking extensive home improvements are a considerable step up from those required in acquiring and using second-hand clothes or furnishings, and further still from those required to recycle packaging. Yet some householders actively pursue these materials regardless, and by doing so are willing to complicate their home improvement projects. Further, rather than pursuing recycled materials because they already have the necessary competencies and are familiar with the systems involved, most appear to acquire them along the way (that is by doing), as concluded by Goodsell (2008).

Discussion

In this section we return to the questions we posed at the start of this chapter, which provide the structure for the following discussion and concluding remarks:

1. As householders undertake food provision, furnishing, and home improvement
and maintenance, what skills, spatial and temporal characteristics, and systems
of provision are involved when second-hand goods and materials are substituted
for new?

A summary of our exploration of themes and household domains is presented in Table 6.1. Across the domains of practice we find that, while skills and knowledge are important enablers of practices that incorporate second-hand goods, these are not the only predictors of the use of second-hand goods. Similarly, while planning, time, storage and space are also important factors in determining the accomplishment of household practices that use second-hand goods and materials, some people 'make' time and space, even when they are just as constrained as others

Table 6.1   Summary of Findings Across Domains of Household Practice and Themes

Domain of practice	Materials or goods of concern	Required competencies	Spatial/ temporal factors	Systems of provision	
				Informal	Formal
Food and beverage provision	Packaging and food waste	MINIMAL: Collect recyclable material; preparation; deposit in appropriate bin; place bin on kerbside OR Collect recyclable material; preparation; locate depots; arrange transport and plan trip to deposit	Provision of own interim container; collection bin provided by council; requires placing on kerb by householder OR Provision of own interim and collection container; time to transport to depot as required	Interim collection bins for packaging and food waste; compost for food scraps; recycling of material not able to be processed through formal system (plastic bags)	Local government recycling service (collection, sorting, reuse/disposal); OR Recycling depots (sorting, reuse/disposal), householders paid for depositing materials
Home furnishing	Acquisition, use and divestment of used furniture	MEDIUM: locate sources of goods or material; plan trips for acquisition or disposal; acquire or dispose of goods or material (purchase, sale or gifting); clean, repair or reuse	Acquisition, use, divestment, washing/cleaning, repairing, storage	Exchange with friends, family, neighbours, purchase or divestment through garage sales .	Purchase from, sale or donation to, charity or second hand shops, 'car boot' sales, online auction sites
Home maintenance and improvement	Acquisition and use of second-hand building materials and appliances	EXTENSIVE: Plan project; prepare site; research, seek advice; locate and engage building professionals/ tradespeople; locate and acquire tools, products, appliances and materials; supervise/direct work OR undertake work; seek necessary approvals; re-occupy and use	Household accommodation/ temporary relocation; acquisition and storage of materials, products, appliances; preparation of second hand material, appliances; installation and application; use	Family, friends, neighbours to provide advice, skills, materials, labour, support; scavenge products and material from garage sales, hard rubbish collection etc. divestment through family, friends, neighbours	Building professionals provide advice, skills, materials, labour; purchase of products, materials, appliances from suppliers; divestment of recyclable products, materials through online auction sites.; formal planning approval; certification and legality

are in this respect. Based on the cross-comparisons of the three cohort studies, we find that those systems in individual households that require more effort to maintain are associated with significant engagement, expertise, investment of time and resources, and ultimately 'ownership' of the perceived consumption problem. While formative experiences do seem to influence how some householders become active, competent and accomplished recyclers, we also find that both formal and informal social networks are critical parts of the combination.

The formal recycling system for food packaging co-exists alongside informal systems at the individual household scale that may include, for example, composting of food waste or scavenging from hard rubbish. Indeed informal systems seemed to encroach on and fit around every formal system encountered in the study. In particular, 'friends and family' networks appear to be important in complementing formal systems of second-hand exchange, enabling recyclers to find out about formal sources of materials and products, new versions of which may be more readily available in well-publicised retail outlets. It is tempting to speculate that households composed of migrants or itinerant young adults find it more difficult to access informal systems. However our cohorts, consisting mainly of recycling innovators, did not reflect this and appeared to find their way around informal systems relatively quickly. In so doing they invariably made reference to social justice or environmental discourses as motivating factors in seeking informal routes for acquisition and divestment of second-hand goods.

Beyond recognising the co-existence of formal and informal systems, we note that the existence of formal systems precludes and shapes the constraints within which household-scale innovations take place – and may conceivably shape consumption as a result. For example a well-organised kerbside waste collection scheme allows for packaging waste to grow since there is little requirement on households to manage this beyond bin size and frequency of collection. The existence of the system therefore *enables* higher throughput of packaging materials in the household. Since recycling processes create significant attendant energy, with environmental and economic costs, the higher throughput of packaging is not necessarily environmentally desirable. Furthermore these systems *enable* widespread proliferation and market domination of supermarkets, which are dependent on energy intensive packaging and storage requirements (for example freezers) to provide and store food. In turn this system effectively *reduces* the choices available to consumers who may prefer to source their food and drink from alternative, small-scale, less packaging intensive suppliers.

Where activities are undertaken within a largely informal system, levels of knowledge and agreement over what constitutes 'correct' recycling are relatively low. Hence formal systems are important in setting 'standards' of practice. Our concern here is that the 'wrong' systems may allow or even encourage practices which lead overall to higher levels of consumption. Thus recycling potentially legitimises consumption and, rather than being seen in itself as 'good', needs to be examined at a systemic level.

2. What factors assist or hinder the reuse or recycling of goods and materials at the household scale and how can this knowledge inform new policy approaches aimed at facilitating more widespread reuse practices?

Referring back to the rationalisation discourse of sustainable consumption, it could be argued that households seeking to use recycled goods and materials in their homes are driven by budgetary constraints or cost saving (that is, rationalisation of their actions). However we found several alternative discourses running through all three data sets. Even households in Cohort 1 that carried out retro-fitting primarily for the stated reason of reducing household running costs (energy and water) were not driven by monetary factors alone; they overlaid this with a strong environmental ethic and were not only concerned about human environmental impact but were equally concerned about intergenerational equity. Further, some of their future plans for solar power or hot water appeared to be predominantly driven by status, or the appearance of doing the 'right thing', rather than by economic or genuine environmental discourses. Household members in Cohort 2 who were active recyclers and users of second-hand goods subscribed to an efficiency discourse in which they were uncomfortable about 'wasting' goods that still had use value. The households in Cohort 3 apparently follow several discourses, but economic rationalisation was not one of them. In fact this discourse was openly rejected and instead this cohort was found to align its household's 'green' home improvement with social and environmental justice causes.

Given this, policy approaches which employ rationalisation discourses, using rebates and other market mechanisms, can only be expected to be partially successful at best. While further research is required to build on the three cohort studies presented here, we conclude that current and future policy making will need to openly address alternative discourses if second-hand goods and materials are to more successfully infiltrate households. Further, policy makers will need to acknowledge and respond to:

- The existence of informal exchange networks of knowledge, skills and resources, which are important determinants of participation in recycling/ reuse. Use of these requires access to competent, clear advice regarding environmental performance/options.
- The reality that householders can operate within sophisticated, often systemic frames rather than individual, economic maximising frames. Programs should consequently be designed with this in mind and draw on social and environmental justice discourses.
- The configuration of systems of provision requires further research, especially in regard to relations between formal and informal systems.

A subtle policy approach is required to increase the volume and type of second-hand goods and materials being substituted for new ones in various domains within the household. Rather than focus only on formal systems, any such approach needs to recognise and support householder agency within existing formal systems and support the sub-systems that operate at varying levels of formality. One such approach is adaptive co-management. Used extensively in natural resource management (largely in rural environments), co-management is advocated as a successful formula for engaging landowners and government agencies in sharing knowledge and resources (Pretty and Ward 2001, Carlsson and Berkes 2005). Co-management is based on the sharing of rights, responsibilities and power between different scales and sectors of government and the public (Huitema et al. 2009). However with the addition of adaptive features such as learning, experimentation and flexibility, 'co-management' has largely evolved into 'adaptive (co)-management' (Huitema et al. 2009).

Such an approach holds promise, we argue, as a means of supporting community-based innovators in second-hand goods, possibly leading to a transition in the use of second-hand goods in households. Niche innovations might possibly, as part of a set of changing system conditions, be taken up more widely and eventually be incorporated within the socio-technical landscape (Geels 2002). Various models for and explanations of socio-technical transitions (for example Geels and Schot 2007) indicate diverse interpretations and pathways, suggesting that we should be wary in attributing causation to any one particular agency. Nevertheless co-management is not antithetical to ideas of socio-technical transitions, and we suggest that co-management at the household level could be a useful component of environmental policy making. One merit of the co-management approach is that it is less instructive towards the household than traditional regulations or policy edicts and is therefore more suitable for supporting sustainable consumption practices in privately owned dwellings. Regardless of the approach chosen, any policy intervention into household material flows and the systems they encounter, either formal or informal, will need to be supportive rather than overriding. In other words, policy and programs should not replace but rather supplement and enhance existing systems. This will require a considered approach and an in-depth understanding of the interconnectedness of systems of varying formality, and, where appropriate, establishing supporting structures.

Conclusions

Examining material flows through households – specifically those related to second-hand goods and materials – reveals the limits of current policy approaches to landfill diversion as an attempt to reduce waste and encourage recycling and reuse. In households a bewilderingly impressive array of skills and commitment is apparent in the way people go about recycling, reusing and adapting goods and materials to everyday uses. Many of the means of acquisition, stewardship and divestment rely on informal systems of provisioning. The time and effort devoted to activities is difficult to predict from case to case, appearing to involve a mixture of judgements made by participants, as well as views about how things are and 'should' be.

Daily practices such as recycling beverage containers and food waste appear to involve fewer competencies than the other domains of practice. However they may involve considerable effort and innovation and require designated spaces within the home. By contrast, more episodic activities like large-scale renovations require significant competencies and expertise. Yet, again, despite this, renovators are typically up to the task, with some even eschewing professionals and learning by doing, going to great lengths to incorporate second-hand goods and materials in the process.

Policies to foster a transition towards more widespread use of second-hand goods and materials in place of new materials should focus on measures to support existing networks and household-scale innovations. This is a more promising starting point than designing new second-hand systems based on economic incentives. While we have not set out to detect changes in practices or determine the prerequisites for broad scale transition in the uptake of second-hand goods, our work suggests that policies based on discourses of economic rationalisation are likely to fail, as are those which do not recognise the diversity and prevalence of informal systems of provision in practices surrounding second-hand goods. A more productive way forward is to adapt ideas of co-management to household practices involving second-hand goods.

References

Alexander, C., Smaje, C., Curran, A. and Williams, I. 2009a. Evaluation of bulky waste and furniture re-use schemes in England. *Proceedings of the ICE – Waste and Resource Management*, 162(3), 141–150.

Alexander, C., Smaje, C., Timlet, R. and Williams, I. 2009b. Improving social technologies for recycling. *Proceedings of the ICE – Waste and Resource Management*, 162(1), 15–28.

Allon, F. 2008. *Renovation Nation: Our Obsession with Home*. Sydney: UNSW Press.

Attfield, J. 2000. *Wild Things: The Material Culture of Everyday Life*. Oxford: Berg.

Australian Bureau of Statistics. 2006. *Australia's Environment: Issues and Trends*. Catalogue No. 4613.0. Canberra: Australian Bureau of Statistics.

Australian Bureau of Statistics. 2008. *2008 Year Book Australia*. Canberra: Australian Bureau of Statistics.

Baum, S. and Hassan, R. 1999. Home owners, home renovation and residential mobility. *Journal of Sociology*, 35(1), 23–41.

Bourdieu, P. 1990. Structures, habitus, practices, in *Contemporary Social Theory*, edited by A. Elliot. United States: Blackwell Publishers.

Carlsson, L. and Berkes, F. 2005. Co-management: concepts and methodological implications. *Journal of Environmental Management*, 75(1), 65–76.

Chappells, H., Klintman, M., Linden, A.-L., Shove, E., Spaargaren, G. and Van Vliet, B. 2000. *Final Domus Report – Domestic Consumption Utility Services and the Environment.* Lancaster: Universities of Lancaster, Wageningen and Lund.

Cooper, T. 2000. Slower consumption reflections on product life spans and the 'throwaway society'. *Journal of Industrial Ecology,* 9(1–2), 51–67.

Darier, E. 1996. The politics and power effects of garbage recycling in Halifax, Canada. *Local Environment,* 1(1), 63–86.

European Commission. 2008. *Attitudes of European Citizens Towards the Environment.* Brussels: European Commission.

Geels, F. 2002. Technological transitions as evolutionary configuration processes: a multi-level perspective and a case study. *Research Policy,* 31(8–9), 1257–1274.

Geels, F. and Schot, J. 2007. Typology of sociotechnical transition pathways. *Research Policy* 36(3), 399–417.

Giddens, A. 1984. *The Constitution of Society: Outline of the Theory of Structuration.* Cambridge: Polity Press.

Goodsell, T.L. 2008. Diluting the cesspool: families, home improvement, and social change. *Journal of Family Issues,* 29(4), 539–565.

Gram-Hanssen, K. and Bech-Danielsen, C. 2004. House, home and identity from a consumption perspective. *Housing, Theory and Society,* 21(1), 17–26.

Gregson, N. 2007. *Living with Things: Ridding, Accommodation, Dwelling.* Wantage: Sean Kingston Publishing.

Gregson, N. and Crewe, L. 2003. *Second-Hand Cultures.* Oxford: Berg.

Gregson, N., Metcalfe, A. and Crewe, L. 2007a. Identity, mobility and the throwaway society. *Environment and Planning D: Society and Space* 25(4), 682–700.

Gregson, N., Metcalfe, A. and Crewe, L. 2007b. Moving things along: the conduits and practices of divestment in consumption. *Transactions of the Institute of British Geographers,* 32(2), 187–200.

Gregson, N., Metcalfe, A. and Crewe, L. 2009. Practices of object maintenance and repair: how consumers attend to consumer objects within the home. *Journal of Consumer Culture,* 9(2), 248–272.

Hand, M., Shove, E. and Southerton, D. 2007. Home extensions in the United Kingdom: space, time, and practice. *Environment and Planning D: Society and Space,* 25(4), 668–681.

Hawkins, G. 2006. *The Ethics of Waste: How We Relate to Rubbish.* Sydney: University of New South Wales Press.

Hobson, K. 2002. Competing discourses of sustainable consumption: does the 'rationalisation' of lifestyles make sense? *Environmental Politics,* 11(2), 95–120.

Hobson, K. (2006). Bins, bulbs, and shower timers: on the 'techno-ethics' of sustainable living. *Ethics, Place & Environment,* 9(3), 317–336.

Huitema, D., Erik, M., Egas, W., Moellenkamp, S., Pahl-Wostl, C. and Yalcin, R. 2009. Adaptive water governance: assessing the institutional prescriptions of adaptive (co-)management from a governance perspective and defining a research agenda. *Ecology and Society* [Online], 14(1), Art: 26. Available at: http://www.ecologyandsociety.org/vol14/iss1/art26/ [accessed 15 September 2010].

Lucas, G. 2002. Disposability and dispossession in the twentieth century. *Journal of Material Culture*, 7(1), 5–22.

Miller, D. 1987. *Material Culture and Mass Consumption*. Oxford: Blackwell.

Miller, D. 1998. *A Theory of Shopping*. Cambridge: Polity Press.

Miller, D. 2001. *Home Possessions: Material Culture behind Closed Doors*. Oxford, New York: Berg.

Pretty, J. and Ward, H. 2001. Social capital and the environment. *World Development* 29(2), 209–227.

Reckwitz, A. 2002a. The status of the 'material' in theories of culture: from 'social structure' to 'artefacts'. *Journal for the Theory of Social Behaviour*, 32(2), 195–217.

Reckwitz, A. 2002b. Toward a theory of social practices: a development in culturalist theorizing. *European Journal of Social Theory*, 5(2), 243–263.

Røpke, I. 2009. Theories of practice – new inspiration for ecological economic studies on consumption. *Ecological Economics*, 68(10), 2490–2497.

Schatzki, T.R. 2002. *The Site of the Social: A Philosophical Account of Social Life and Change*. University Park, Pennsylvania: Penn State University Press.

Shove, E. 2003. *Comfort, Cleanliness and Convenience – the Social Organization of Normality*. Oxford: Berg.

Shove, E. and Southerton, D. 2000. Defrosting the freezer: from novelty to convenience: a narrative of normalization. *Journal of Material Culture*, 5(3), 301–319.

Shove, E. and Hand, M. 2005. *The Restless Kitchen: Possession, Performance and Renewal*. Kitchens and Bathrooms: Changing Technologies, Practices and Social Organisation – Implications for Sustainability, Interdisciplinary Workshop, University of Manchester, United Kingdom, 27–28 January.

Shove, E., Watson, M., Hand, M. and Ingram, J. 2007. *The Design of Everyday Life*. Oxford: Berg.

Valentine, G. 2006. Relevance and rigour: the advantages of reusing and scaling up qualitative data. *Environment and Planning A*, 38(3), 413–415.

Shove, E., Chappells, H., Lutzenhiser, L. and Hackett, B. 2008. Comfort in a lower carbon society. *Building Research & Information*, 36(4), 307–311.

Warde, A. 2005. Consumption and theories of practice. *Journal of Consumer Culture*, 5(2), 131–153.

Watson, M. and Shove, E. 2008. Product, competence, project and practice. *Journal of Consumer Culture*, 8(1), 69–89.

Chapter 7

Bottled Water Practices: Reconfiguring Drinking in Bangkok Households

Gay Hawkins and Kane Race

Introduction

This chapter examines bottled water practices in Bangkok: how they function practically; how they become meaningful and normalised; and how they interact with everyday household water routines. Single-serve bottled water is generally classified as a fast food commodity driven by the logics of the global beverage industry and designed to be consumed outside the house. Drinking water from a branded bottle is seen as a form of leisure consumption vastly different from turning on the tap and interacting with a state utility. But can bottles and taps be so easily set in opposition, one emblematic of a product, the other of a service? Is the distinction between these two drinking practices as clear-cut as this? What of the many places where state or commercial water utilities are non-existent, underdeveloped or unsafe? In these settings the meanings and efficacy of bottled water, bought from street vendors or home delivered, operate far beyond the registers of frivolous leisure consumption. This is just one of many examples that blur the distinction between taps and bottles and reveal the complexity of drinking water practices. Both bottles and taps deliver water and discipline its consumption via a variety of socio-technical and economic arrangements. And in many settings these different arrangements are inter-articulated, in the sense that they influence and interact with each other in complex and various ways. The challenge is to understand these interactions and to investigate the processes whereby drinking practices are made meaningful.

Our interest is in how the matter of the plastic bottle comes to matter in different settings. How does a fully materialised account of drinking practices make it possible to think about the interactions between bottles and taps in more productive ways? A focus on drinking practices foregrounds the efficacy of bottles in different settings. It also shows how objects and practices are mutually constitutive. This approach situates the water bottle within the routines and habits of everyday life *and* the ways in which artefacts participate in these routines and help constitute the social. Practices, then, are always more-than-human. Rather than see them as an expression of human agency or culture they have to be understood as complex associations of materials, technologies, norms and bodily habits that are sustained and modified through repeated performance or enactment. In the case of the plastic water bottle, these practices vary significantly according to context. As an object designed for portability and single use It is most alive outside the home. How, then, does the bottle mediate inside and outside, or stasis and mobility? And how does it impact on household water practices? In what ways does the tap as the endpoint of a service interact with the bottle as beverage or commodity? How do these distinctions reverberate on the 'economy of qualities' (Callon et al. 2002) that variously values water? These are the larger questions driving this chapter, but first we explain our approach to the question of practice.

Thinking Practices

There is no question that drinking water from plastic bottles is a relatively new practice. Think back 20 years and ponder where you saw water bottles: in the gym, definitely, on desks, maybe. Now they are ubiquitous. Everywhere you look there are bottles: in cars, tucked into special pockets in backpacks, on lecture theatre desks, in hands as people jog, handed out free at events, lying in the gutter discarded; the list goes on. Bottles have become part of the material density of everyday life; they have become domesticated, and in the process they have inaugurated a range of new conducts. The challenge is to understand how this object and the activities that have incorporated it into daily use have co-evolved: how bottles and bodies have become connected and interact, generating new ontological realities for drinking (Mol 2003: 6).

Studies of consumption or material culture are of limited help. As Elizabeth Shove and colleagues (2007), Bill Brown (2003), Jane Bennett (2001) and others have argued these approaches pay insufficient attention to the materiality of the commodity: to the ways in which its material qualities participate in the constitution of markets and consumers. Theories of material culture implicitly render the commodity a passive object of cultural inscription, a surface on which 'culture' gets to work and makes meaning. In these frameworks practices are largely things humans do *to* or *with* things to express identity, social positioning or deeper social order. While there might be a commitment to the social as constructed, 'practices' in these frameworks appear relentlessly human – as emanating from human consciousness, intentionality and discourse.

What is missing in these approaches is an understanding of how material things participate in shaping bodies, actions and meanings; what uses they afford; and how these affordances are continually extended in practice,[1] which is to say how the social is both produced and practised in and through relations with artefacts and is therefore not exterior to these relations. Hence the turn to science and technology studies (STS) in many accounts of practice, for it is here that distributed forms of agency are recognised, and where the more-than-human or hybrid dynamics of meaning and matter are central. Also important in this approach is the refusal to

---

1   For an excellent account of 'affordance' see Mike Michael (2000: 61–67).

allow macro categories such as 'culture', 'economy', 'society' any 'trans-historical ontological validity', as Tony Bennett puts it (2009: 102). What this means is that, in this analytic mode, reality is enacted or performed through the multiple relations whereby things get associated. So it is not a matter of identifying actions and practices as evidence of social forces or representations of deeper structures but rather of tracing how 'the social' emerges in the dynamics of both durable and fleeting assemblages.

However as Shove and colleagues (2007) argue, STS still has limitations when it comes to developing a fully materialised account of practices and enactment: 'The Latourian contention that artefacts literally construct socialness has yet to be worked through in any detail' (2007: 14). *The Design of Everyday Life* is their attempt to redress this problem with a close analysis of how various materials become implicated in everyday practices of consumption, renovating, design, photography and more. The empirical focus is on tracing the relations between objects, bodies, meanings, forms of competence and routines in a range of settings. What is especially valuable about this method is the commitment to understanding how material things and technologies become integrated into practices as performance, and how this both realises their various material affordances and also, at the same time, stabilises social relations over time (2007: 148). This does not mean that practices become fixed performances endlessly repeated. Practices continually evolve and are continually integrating new elements that might emerge from bodies – material that presents new possibilities, practical knowledge that shifts over time or unexpected disruptions to routines.

Drawing on Theodore Schatzki (1996), the key point Shove and colleagues make is the distinction between practice-as-performance and practice-as-entity. Practice-as-entity has a relatively enduring existence over time and space. It refers to the ways in which practices are made durable via the relationships between norms, materials, shared meanings and bodily routines. Practice-as-performance is the specific enactment, the active doing through which practice-as-entity is sustained, reproduced and changed. This refers to the contingent dimension of practices, the ways in which practices are both reproduced and continually reinvented in action (Shove et al. 2007: 12–13). This theorisation of practice pushes accounts of domestication and appropriation of technology and objects beyond the registers of mutual co-evolution of people and things. This can implicitly endorse an approach in which humans tame things: in which the thing is incorporated into existing routines and spaces rather than being actively involved in making and remaking them. In other words, it can imply a certain form of material stasis, or 'socio-technical closure'. Once the thing is stabilised or embedded in contexts it remains relatively unchanged. The value of Shove and colleagues' approach is its insistence on the role of materials as actants that can suggest and transform practices; that is, on practices as complex assemblages of the human and non-human that are always on the move (Shove et al. 2007: 8).

In our wish to make sense of the massive growth in bottled water consumption this account of practices is extremely valuable. It forces attention on 'drinking' as both a practice-as-entity and practice-as-performance, and it foregrounds the role of the bottle as a commodity that participates in the emergence of new practices. This also opens up a new approach to thinking about sustainability. Rather than critique the rise of the bottle as an environmental catastrophe, or as a threat to the provision of safe drinking water, (which is the dominant trope in most analyses of bottled water to date) a focus on drinking practices pays close attention to the ontological realities of bottles in action. The issue then becomes how does drinking water *from bottles* emerge as a new practice? What is involved in this practice: what practical knowledges, routines, norms and more sustain it? And what kinds of implications does bottled water practice-as-performance have for other drinking practices? If drinking water from taps can be considered a practice-as-entity, a spatially and temporally enduring assemblage that frames drinking water as a service, how do bottles interact and interfere with this? How have bottles reinvented water, its technical delivery and ways of drinking?

These questions have a different political orientation to that of critique. They involve an *ontological* politics because they focus on how bottles come to matter – on the kinds of worlds they perform. These worlds or realities are not simply 'constructed', they have to be practised or enacted into being and this involves choices, obstructions and interference from other ontological realities. While we are concerned to trace bottled water practices in a range of settings we are also interested in how these might *interfere* with other sorts of drinking practices. 'Interference' here is an STS term that in John Law's (2004: 5) account means that realities are being practised everywhere, that they are complex, uncertain and interact with each other – this is difference. This difference suggests how ontological realities may become ontological politics because difference can mean both conflict and dissent and the imagination of alternative realities. As Annemarie Mol (2003: 7) says: '[I]f reality is multiple it is also political.'

This account of politics extends understandings of practices in important ways. For a start it makes it possible to see how bottled water enacts ontological realities that are different from and may (or may not) interfere with other drinking practices. Reducing these differences to the logics of a commodity/service opposition does not get anywhere near understanding how bottles and taps interact. This opposition denies the way drinking practices might mix up the ontologies of taps and bottles rather than oppose them, or how they might get enacted in specific places, sometimes producing interference and at other times innovation.

To see these ontological politics in action we turn now to a case study of bottles in action in Bangkok. Our aim is to document a range of provisional arrangements that are in place for drinking water. These involve all manner of objects, routines and networks in households and beyond, and they show how bottles are generating new practices, markets and drinking performances. They are also being incorporated into existing practice-as-entity regimes in ways that complicate the distinction between product and service. The key purpose of this case study is to investigate how specific material affordances of plastic bottles, in all their varieties, are realised in practice. And how, in some arrangements, these practices enact ontological realities that interfere with the imagination and provision of more sustainable alternatives.

Provisional Networks of Water in Bangkok

The process of organising everyday drinking water involves collaboration with a range of human and non-human others in the household and beyond. In Thailand, where we conducted fieldwork, rainwater was traditionally collected in large earthen jars placed outside households – a practice that persists in some parts of the country today. More recently houses have been fitted with galvanised drainage gutters and metal pipes directing the water into big, ceramic storage jars. In urban Bangkok, where atmospheric contamination of rainwater has become a problem, many households and condominium residents inevitably enter into some sort of arrangement with the metropolitan tap water system. Vestiges of past practices are nevertheless apparent in the accounts of some of our Bangkok informants, who specified the taste of rain as a value they loved in drinking water.

Since 1999 the Metropolitan Waterworks Authority (MWA), a state enterprise administered by the Ministry of Interior, has guaranteed that Bangkok tap water complies with WHO drinking water standards after treatment at the source. Water is regularly tested for key contaminants at various locations in the municipal network and the results are uploaded onto a consumer map on the MWA website. Most of our informants were aware of state guarantees, but expressed doubts about the quality of pipes and their maintenance, especially in older buildings and sections of the city. This view of pipes as an unreliable intermediary was also promoted by bottled water industry informants we interviewed:

> *No matter how good your water is, somehow you have to find a way to deliver the water to the consumer and when you talk about tap water, you are talking about the piping system, and the piping ages – rust, pollution, all that sort of thing. People doubt the maintenance.*

In these circumstances of perceived unreliability or risk, householders described entering into a variety of arrangements with mundane devices and technologies in the interests of organising safe drinking practices. These included boiling water from the tap, installing filtering systems in kitchens, arranging home delivery of drinking water (Home and Office Delivery, or 'HOD') or making use of the water vending machines which can be found in some Bangkok neighbourhoods. Water vending machines began appearing from the late 1990s. These machines were marketed as a low cost source of safe drinking water. Essentially a filtering device, they are connected to the municipal water system and run by private operators. They are coin operated and require the consumer to bring their own container. More recently large PET bottles bought from supermarkets have emerged as another domestic water source. Single serve PET bottles are generally not seen as a staple form of household consumption; they are usually bought while out and about. They may, however, find their way into the house at the end of a daily excursion where they are often refilled and reused. Each of these arrangements involves varying degrees of labour inside and outside the household, and the coordination of bodies, technologies, forms of competence and routine. Such routines are themselves entangled with other routines – preparing meals, going shopping, stocking the kitchen – and must be adjusted in relation to the material particularities and conditions of the elements at hand: water is heavy, transporting sufficient volumes for household use calls for a vehicle, Bangkok traffic is bad, the water vending machine is hard to get to and so on. What became apparent in our study was the multiplicity of these practices and their adaptability to different conditions and elements over time. In the context of the arrangement of new drinking water practices, tap water tended to be limited to other domestic uses such as washing and bathing, cleaning kitchenware and clothes, and cooking.

One way of approaching the multiplicity of drinking water practices is to position the consumer as a participant in *provisional networks* of distribution, preparation and supply. In the household these networks are provisional in two senses: they are about *making something available*, and they involve routinised practices which may nevertheless be subject to revision and some degree of change or innovation in the presence of new technologies, services or products such as filters, water vending machines or PET bottles.[2] This is the nature of the relation between practice-as-entity and practice-as-performance. Drinking practices are also subject to change in the context of new styles of life, such as condominium living, a major and relatively recent development in Bangkok which, alongside convenience stores, supermarkets and shopping centres, is promoted as a hallmark of cosmopolitan modernity. Our informants were engaged, or had been engaged at particular points in time, in degrees of reflexive activity about the various components of such networks in relation to everyday values like reliability, cost, health risks, taste, social status, practicality and convenience. This shows that it is not simply new technologies or products that inaugurate changes in practices but rather their capacity to prompt reflexive activity and generate what we might call 'ontological doubt' about the security or stability of existing arrangements.

Good or Service?

Is the provision of drinking water a good or a service? Framed as a service, it could be thought to follow the logic of what Michel Callon and colleagues describe as 'making available', where the customer, 'by opening a tap ... sets in motion

---

2   This metaphor of 'making available' is particularly apt in the Thai context, as water is intimately tied into practices of hospitality. The first thing houseguests customarily receive on entering someone else's house is water. A common expression for generosity is *nam jai*, literally, to have a water heart.

a complex arrangement of humans and non-humans whose actions have been adjusted in relation to one another and prepared for mobilization at any time and at any point of access to the network' (2002: 208). Services imply an ongoing relation with the customer, in which the provider agrees to make certain things available on certain conditions for a period of time. Moreover in service provision, customers become 'an element in the system of action. They act, react and most importantly interact' (209) – a feature that Callon and colleagues see as producing a customisation of the relation. In some ways Callon and colleagues seek to blur the distinction between goods and services. Goods are cast in terms of the sequence of actions and operations in which their properties are worked on and qualified. Here products are goods seen from the point of view of their production, consumption and circulation, a process that involves various forms of organisation and reflexive activity on the part of economic agents. Because these processes include specific devices for registering and incorporating consumer desires and preferences, the distinction between product and service becomes blurred. The general aim seems to be to direct attention to the contestable processes through which products are qualified and their properties stabilised.

Importantly for Callon and his co-authors, the qualification of a product can consist of work on the image of the product, or work on its actual material form, or both. There is no need to distinguish between 'primary' and 'secondary' qualities in this regard: for the purpose of positioning, these attributes share the same ontological status. This point is particularly helpful for thinking about the bottled water market, and a welcome rejoinder to the popular notion of bottled water as a case of constructing semiotic desires about nothing. For nothing is learnt in such approaches about how the bottle itself comes to matter.

This blurring of the distinction between goods and services is extremely pertinent to drinking water practices in Bangkok. It offers a valuable framework for thinking about how the properties of *both* are continually modified in interactions with the economic agent (or householder) and other elements of drinking water practices. These interactions reveal, as Callon and colleagues argue, the contestable processes through which products and services are qualified and their properties stabilised. In the case of water these processes most often involve its qualification as safe or drinkable. As we have seen, home consumption of drinking water from non-tap or MWA sources is largely dominated by the HOD industry and water from neighbourhood vending machines, though use of PET bottles is on the rise. The HOD industry is often regarded as the environmentally friendly approach to dispensing bottled water because it reuses the large 19 litre PC bottles in which it delivers the product. For this reason it is understood within the industry as a service:

> The bottled water cooler industry differs from the bottled water industry, which bottles water in small bottles, in that the former is service-based whereas the latter is product-based. This difference is very significant for the organisational structure of the bottled water cooler company (Barnett 2008: 30).

Specifically, the company must arrange regular scheduled delivery and collection, maintenance and repair of the relevant technologies, customer care, and industrial cleaning and reuse of the 19 litre PC (polycarbonate) bottles which typify the industry. Cleaning the PC bottles is a complex operation that involves administering cleaning solution and rinsing the inside and outside of the bottles using a jetspray system. As one HOD industry informant explained, it must also contend with other possibilities:

> When your bottle comes from the market, you never know what your bottle has been in contact with, what kind of contamination. That's why our company tried to persuade the consumer to recap your bottles after use, but we haven't been successful as yet, because people say, what the hell? You know, so the contamination comes along.
>
> The worst part is when your customer or your consumer doesn't really, you know, care about, take care of the bottle. They put it as a container for some kind of petroleum, or some kind of pesticides, or whatever, and when you bring this bottle back, the only solution for that is that you have to discard the bottles. So when the bottle comes back from the market the first thing that the operator does is, they have to do the sniffing test. And if it smells of petroleum or something then they put it aside. The majority of times you have to destroy the bottle. And then they do the cleaning of the outside, which is quite a difficult one, and after that, you do the cleaning of the inside.

Here, the customer is positioned as an unreliable intermediary. We can see here that the company's reuse of the bottle requires it to account for the customer–bottle relation. In other words, the reusability of the bottle constitutes certain relations between providers, customers and bottles, rendering each an active element in the provisional network.

In order to think about how disposability is constituted in relation to the bottle, it may be useful to consider another situation in which the bottle is reused by the producer. Apart from the HOD industry, an example from the Thai context is the use of single-serve reusable glass bottles to sell soft drinks. (In Thailand, traditional trade channels actually prefer this option because of the higher margins involved.) We can see that in both these cases, *reuse of the bottle* makes certain demands on companies and distributors, ranging from storage space, in the case of empty glass bottles (which makes modern and convenience stores reluctant to take up this option), to the pickup and industrial cleaning of used bottles, in the case of HOD services. Moreover the premium on the bottle's *reusability* grounds the bottle in certain ways. If the bottle travels too far, the distributor makes a loss – thus the mobility of the bottle must be accounted for. This is achieved by grounding consumption in certain locations (such as the home/office in the case of HOD services); by introducing a deposit/refund system; by requiring consumption on the spot (for example in restaurants); or by other inventive means (soft drink vendors working with glass bottles in markets often transfer the contents into a plastic bag, add a straw and ice, and fasten the bag at the top with a rubber band, allowing the vendor to retain the bottle while leaving the consumer free to roam).

In these instances we can see how a property of the bottle (reusability) institutes specific relations and practices and necessitates certain socio-technical arrangements on the part of suppliers, distributors and consumers. These examples also help to reveal some of the specific affordances of the disposable PET plastic bottle. In quite material and practical ways, the single-serve PET bottle helps to equip or furnish the mobile consumer. Disposability allows a severing of the relation between the provider and the consumer. Indeed the single-serve PET bottle was most often associated with consumption on-the-move rather than the household, the main appeal being its 'convenience'. But what can we say about the growing use of PET bottles inside the home? Since mobility is not a salient value in this context, how does this product position itself in relation to the alternatives?

Not surprisingly it is precisely the material components in HOD networks of delivery and supply – and their tendency to fail – that the PET bottled water industry exploits in its attempts to penetrate the household market. As the head of bottled water at Nestle Thailand reported in the industry magazine:

> The new generation of consumers are health-conscious so they started questioning the cleanliness of the returnable 19 litre bottles, especially when the bottle condition and label looked old, alerting a concern on the potentials of poor washing and disinfection. (Muernmart 2008: 8)

Here the company is exploiting consumer experiences of HOD bottles, in particular the mishaps associated with the reuse of bottles in given networks of supply. In other words the property of disposability acquires value in relation to bad experiences with the reusable bottles that characterise the household market. Calculations such as these prompted Nestle to introduce a new 6 litre PET bottle into the convenience store trade channel. But what is especially interesting here is that the positioning of bottled water within this market involves specification, not only of the symbolic, but also the material properties of the product. Consider for example the material-practical concerns that are cited to position it:

> Convenient to buy, not that heavy, suitable to carry back home, and a good price per litre – ideal for the new generation family of 2, husband and wife. (Muernmart 2008: 7)

Note in this passage the allusion to a new, 'modernised' consumer with a modern lifestyle, but also the producer's familiarity with the provisional networks within which such consumers participate. Once again, the bottle – including its actual material form – is carefully adjusted to and designed to equip certain routines of modern life. The appeal to new styles of life is not merely symbolic, but promotes the bottle in terms of its specific affordances: not too heavy, but a sufficient quantity of water for a modern household; something you might pick up at the convenience store on your way back to the condominium. Notice also that these material properties gain their value and significance in relation to existing provisional networks – specifically the potential of these networks to generate insecurity about the cleaning of bottles and the quality of the water.

Conclusion

What can we take from this case study of drinking water in Bangkok? Firstly, and perhaps most obviously, we can see close attention paid to consumer practices, routines and provisional arrangements on the part of bottled water companies. Perceptions of tap water were affected by the appearance of new technologies and products which promote themselves in terms of safety and reliability. Most of the provisional arrangements we encountered were driven, at some level, by concerns about safety, though it was often difficult to distinguish these from concerns around social distinction, taste and convenience. Here ontological doubt about the safety of different provisional arrangements merged with other rationales for consumption, including social status, taste, convenience and health. In these contexts, tap water was re-purposed to other household activities, such as cleaning, bathing and cooking. In this context these companies are paying close attention to actual drinking arrangements and the ways in which they continually qualify and re-qualify water. The promotional appeal to a modernised lifestyle goes hand in hand with a positioning of products that foregrounds their material properties and affordances as much as their image. Indeed these material dimensions become part of the brand, insofar as the bottle is positioned and experienced in terms of everyday values such as practicality, convenience and cleanliness. If practice theory 'shifts bodily movements, things, practical knowledge and routine to the centre of its vocabulary', it would seem that this is one of the key registers in which producers engage in their attempts to position their products competitively (Reckwitz 2002: 251).

Secondly, the material properties and affordances of the bottle acquire their value relationally. More specifically, they acquire their value in relation to the provisional arrangements to which households are accustomed, and the perceived shortcomings of those arrangements. The quality of disposability, afforded by PET plastic, has no really salient value in the household except *in relation to* some of the undesirable aspects of given provisional networks, such as HOD delivery. The salience of this specific characteristic depends on or contrasts with the performance of other materials against which it is compared, such as reusable PC bottles. In this respect, materials such as disposable PET plastic can be regarded as involved in 'overlapping webs of relational performance', to borrow Shove and colleagues' suggestive phrase (2007). These performances consist of specific applications of given materials, and are themselves 'relative, provisional and inherently precarious' (Shove et al. 2007: 105). Producers seek to make new markets by promoting specific expectations of material – object performance. Meanwhile, the material value of PET plastic takes shape in relation to given performances of PC plastic.

This point connects with the next observation we can draw from the Bangkok case study, which concerns the distinction between products and services. Indeed when it comes to water provision it is tempting to argue that the quality of disposability enacts a distinction between product and service in the home space. If, within the logic of service provision, customers are constituted as active elements within a system of action, we can see that much of the appeal of PET plastic is that it severs the need for continued participation within the network, or any ongoing relation between customer and provider. This value can be understood in terms of mobility (dispensing with the need to return the bottle to the provider) or in terms of dislocation (releasing the customer from a provisional network in which certain elements, such as dirty bottles, are unsatisfactorily controlled or accounted for, thus providing safer or cleaner water). In either case, with a disposable PET bottle the ongoing relation between customer and provider can cease at the point of purchase. A question that arises though in relation to the circulation of a good so basic to human sustenance as water is whether severability from provisional networks (as depicted in the logic of products and services) is an adequate formulation? Or whether the provision of clean drinking water ought to be characterised in other terms entirely: not in terms of the logic of choice, for example, but a logic of care, which Annemarie Mol characterises as an interactive, ongoing and open-ended process that does not stop at the point of transaction but rather requires continual modification depending on results and human need (Mol 2008).

Finally, the multiplicity of drinking water arrangements in Bangkok, and their interference with one another, suggests the need to further theorise *convenience*. Not only is convenience commonly cited to account for consumer preferences, it is frequently cited specifically in relation to bottled water marketing (Ward et al. 2009). Elizabeth Shove has discussed convenience in terms of the ability to shift time (2003). In the case of bottled water we can see that it is also connected to practices of movement and thus has spatial dimensions. Etymologically, convenience implies a coming together. It is a coming together of different elements in a network of humans and non-humans in an arrangement that is adjusted to the routines of key actors in that network. As a demand, it takes its bearings from given routines, procedures and competencies. Just as often, though, it is projected as a property onto specific goods, services or arrangements. Convenient products are those that are well-attuned to stabilised routines and procedures in given relations of affordance. Or they promise a new, more desirable or efficient stabilisation, in which specific forms of labour, cost, or time are redistributed. But what is also apparent from our study of Bangkok drinking practices is that convenience is not a transparent value. As we have discussed, it takes its bearings from given configurations and provisional networks (in what situation is lugging a six litre PET bottle home from the convenience store 'convenient' exactly?). And as well as citing given routines, the notion of convenience also disrupts or supplants them, making certain practices and competencies redundant and creating the need for new forms of labour, cost and routine. This is how we understand Mol and Law's claim that different practical ontologies interfere with one another. The value of the object-oriented approach we are advocating is that it can better take account of the dense materiality of stuff like PET plastic and the subtle ways it works itself into our lives, while also exposing the contingency of these workings, their relationality. This could make emerging/naturalised drinking practices more open to change, which is to say, less essential.

References

Barnett, M. 2008. Bottled Water Coolers. *Asia Middle East Bottled Water Magazine*, December, 21–31.

Bennett, J. 2001. *The Enchantment of Modern Life*. Princeton: Princeton University Press.

Bennett, T. 2009. Museum, field, colony: colonial governmentality and the circulation of reference. *Journal of Cultural Economy*, 2(1), 99–116.

Brown, B. 2003. *A Sense of Things*. Chicago: University of Chicago Press.

Callon, M., Meadel, C. and Rabeharisoa, V. 2002. The economies of qualities. *Economy and Society*, 31(2), 194–217.

Law, J. 2004. Matter-ing: or how STS might contribute? Lancaster: Centre for Science Studies, Lancaster University.

Michael, M. 2000. *Reconnecting Culture, Technology and Nature: From Society to Heterogeneity*. London: Routledge.

Mol, A. 2003. *The Body Multiple: Ontology in Medical Practice*. Durham, London: Duke University Press.

Mol, A. 2008. *The Logic of Care: Health and the Problem of Patient Choice*. London and New York: Routledge.

Muernmart, A. 2008. The 'war' of waters: Nestle vs Singha, *Asian Middle East Bottled Water Magazine*, December, 5–14.

Reckwitz, A. 2002. Toward a theory of social practices: a development in culturalist theorizing. *European Journal of Social Theory*, 5(2), 243–263.

Schatzki, T. 1996. *Social Practices: A Wittgensteinian Approach to Human Activity and the Social*. Cambridge: Cambridge University Press.

Shove, E. 2003. *Comfort, Cleanliness and Convenience: The Social Organization of Normality*. Oxford and New York: Berg.

Shove, E., Watson, M., Hand, M. and Ingram, J. 2007. *The Design of Everyday Life*. Oxford: Berg.

Ward, L., Cain, O., Mullally, R., Holliday, K., Wernham, A., Baillie, P. and Greenfield, S. 2009. Health beliefs about bottled water: a qualitative study. BMC Public Health. [Online]. 9:196doi:10.1186/1471-2458-9-196. Available at: http://www.biomedcentral.com/1471-2458/9/196 [accessed: 15 September 2010].

Discussion: Watch Where That Went – We May Need It Later: Reflections on Material Flows in and through Home

Louise Crabtree

The chapters exploring material flows through domestic spaces shed a fascinating light on the complexities, opportunities and contradictions manifest in practices and systems of household formation and consumption. They highlight areas of convergence, mismatch, overlap and lag between design, behaviour, norms and economics in the forms and products involved in housing spaces and systems. Moreover the chapters reveal the intricate interactions and blurring between products, identity, lifestyle, behaviour and technology. These chapters reveal the diverse ways in which people enact social, economic and environmental imperatives within existing infrastructures, which are in turn products of and factors in myriad social, economic and environmental imperatives.

The spatialities of material flows within the home exist in feedback loops between practice, norms, formations and behaviours. Robyn Dowling and Emma Power trace the flows and discourses of objects in houses in Sydney's outer suburbs, revealing the interweaving, and often contradiction, of social and environmental imperatives underlying these flows. The multiple juxtapositions and separations of new and recycled goods are shaped by ideas of order, spaciousness and cleanliness, frequently with second-hand goods placed in 'messy', informal or peripheral areas and new objects in formal, more public spaces within the home. In the context of the relatively large suburban homes of that case study, such spatial delineation becomes possible. Such homes are built and marketed with a focus on maximal internal space, with ballooning house size seen by many developers, builders and buyers as providing 'more bang for your buck'. These homes are typically built and marketed with multiple internal spaces, such as multiple living, sleeping and entertaining areas for adults and children, and often with home offices. While the internal delineation of spaces frequently corresponded to a delineation between types of objects in the case studies, it was shot through with ideas of family and memory, such that prized or significant second-hand possessions could take pride of place on the one hand, or be strategically placed to enact familial duties when the relationship to the object was more ambivalent.

Robyn Dowling and Emma Power's chapter thus also highlights the role of home as a primary reflection of and constitutive factor in identity. Many householders exhibited spacious senses of self in line with spacious homes, with unease and stress felt in cluttered or cramped environments. Much has been made of the emergence of large, open-plan homes as a reflection of the demise of public spaces and a generalised trend toward privatisation and individualisation via the internalisation of these spaces within private homes. This is not a discussion to be had here, but it is worth noting that a spacious sense of self, open-plan housing and minimalist interior decoration are recent phenomena: witness the Victorian era's preference for clutter and ornamental decoration, with bare rooms considered to be in poor taste. It is also worth noting, however, that open-plan increasingly exists alongside a resurgence of additional rooms for specific uses, such as home theatres, parents' retreats and studies. Whether large suburban homes are the hallmark of privatisation and the death-knell for sustainability or not, it is evident that there are complex and compelling material flows through these homes, shaped by often contradictory or juxtaposed senses of duty, order, cleanliness and family. Meeting the ongoing housing needs of the nuclear family is possibly the strongest theme running through Dowling and Power's case studies, with most recycling of goods focused on the inheritance or gifting of goods through friendship or kinship ties, plus the storage and circulation of clothes and furniture over various life stages, from birth to death and around again. These pathways, spaces and activities overlap with sustainable and unsustainable activities in unintentional ways, driven by the myriad other imperatives shaping and reflecting homemaking, whether familial relationships, economic thrift or contemporary expectations about the form and structure of home. Indeed, these imperatives and activities cannot be considered in isolation and are bound up in the entirety of homemaking practices.

Ralph Horne, Cecily Maller and Ruth Lane's research adds to this, providing additional data on the imperatives driving environmental practices in the home, revealing not only data about what, why and how people reuse or recycle, but also data on the discourses deployed in framing and understanding these practices. Similar to Dowling and Power's research, that of Horne, Maller and Lane reveals the role of social norms, systems of provision/collection/redistribution and individual perceptions in practices of sourcing and using second-hand goods in the home. Perhaps not surprisingly, different discourses emerged in relation to different practices, with 'mainstream' recycling activities the most frequently cited instances of environmental practices, such as using kerbside glass, plastics and paper collection, and backyard composting. This is suggestive in many ways. Firstly, it is evident that the respondents cited in the study pride themselves on the recycling systems they have established, which interact with formal systems such as kerbside systems, established and promoted heavily in Australia to encourage core home-based environmental behaviours over the past two decades. As such, kerbside recycling has become part of what a 'good' home-based consumer/citizen does, and respondents correspondingly placed themselves within a framework of 'good' home behaviour, proudly displaying their competency as developed according to expectations of home-based environmentalism. This focus on the shaping of behaviour and skilling-up of householders according to expectations of environmental behaviour is a crucial insight, as it highlights an internalisation of broader societal interpretations and policies of environmentalism subsequently enacted in and through modifications to homemaking practices. Horne, Maller and Lane rightly highlight that this may in fact act to shut down practices focusing on reduction or reuse and to increase consumption or use of packaging via the absolution apparently provided by recycling. Secondly, it is unclear whether other systems and practices like sourcing second-hand clothes are as widely perceived as environmental behaviours – certainly they have not been promoted as such in Australia – and respondents largely drew on other discourses of homemaking in their reflections on these practises, such as thrift, life stages, cleanliness and memory.

Thirdly, Horne, Maller and Lane highlight and explore the skills, affordances and adaptations made by homemakers in developing and maintaining systems and spaces for reuse and recycling, highlighting the conscious deployment of time, space and energy for these activities, and the often resultant pride. This echoes Dowling and Power's documentation of identity formation through home, adding to it a documentation of bodily enactments of homemaking practices identified as sustainable by their initiators. Light is also thrown on the role of broader processes and structures in shaping home-based practices, whether the ambiguous role of kerbside recycling, the reluctance of builders to work with recycled timbers, or conflicting time economies between renovators, suppliers and builders. It can be seen that environmentally focused home-based practices are contingent, negotiated and shifting, operating at and across multiple spaces and scales.

Fourthly, areas of possible contention between the case studies and methodologies in Horne, Maller and Lane's paper are suggested. The three pilot studies related to respondents with three different levels of self-identified environmental behaviour. Each group was then asked different questions on the basis of self-perceived differences in activity. As a consummate prober, I would love to know how the data might shift if the questioning remained constant. Do individuals who believe themselves to be involved in more environmental behaviours actually take part in more environmental behaviours (however defined) or identify more of the same suite of behaviours as 'environmental'? Given the messiness of perceptions of normality, family, cleanliness and environmentalism highlighted in other chapters, this question will have me following the research with interest. Again, and as noted by Horne, Maller and Lane, dominant discourses and systems of recycling and other behaviours may not in fact align with practices or systems that generate 'better' outcomes. Hence much actual 'environmental' behaviour may be taking place under other guises (such as a personal preference for second-hand goods without awareness of or concern for their environmental performance) and much touted environmentalism may in fact make matters worse – again, witness the potential role of recycling in normalising packaging and refrigeration.

In the last chapter in this part Gay Hawkins and Kane Race trace the intricate interactions and slippage between product, identity, lifestyle, behaviour and technology, teasing out the often ignored dynamic relationships between product design and use, and the urban infrastructures and identities within and through which these are situated and articulated. Their chapter neatly surmises the combined argument of this section: that 'practice-as-performance is the specific enactment, the active doing through which practice-as-entity is sustained, reproduced and changed'. This both recognises the co-construction of artefact and behaviour and offers scope for intervention and action, as this makes 'practices more open to change, which is to say, less essential'.

The differences in the socio-technical infrastructures surrounding the deployment of plastic water bottle in Bangkok are intriguing and significant. A key point is that certain material properties only make sense in particular socio-physical infrastructures – such as the size and portability of water bottles for domestic water consumption in Bangkok as a reflection and facilitation of particular household formations and identities. However as the preceding chapters on household formation and organisation highlight, products and designs are constantly reworked and reinterpreted according to ideas and practices of home, family, self, cleanliness and order, so we could expect that this is not a seamless union – more the start of a dance. Hawkins and Race's research again raises the issue of cleanliness, here overlaid with distrust based on experiences with the failure of mediatory or centralised infrastructures.

These chapters are united in their focus on the fascinating complexity of the materiality of individuals and households engaging with and manipulating the formation of home through managing the often unknown natures and histories of artefacts that reside in or move through the home. In all cases, ideas of normality, family, cleanliness, memory, convenience, lifestyle and economy are bound up in ongoing formations of the behaviours, systems and infrastructure of home.

A core theme here is that calls and efforts for addressing environmental and social justice through domestic systems and practices are always contingent and embedded, and must take heed of this. Further, in taking heed, such endeavours must be based in everyday practices and expectations. So Ash Amin's call for a renewed sense of the commons via infrastructure provision (Amin 2010) may come unstuck when challenged by experiences, histories or perceptions of centralised provision as unsafe, inefficient, unclean or not in step with desirable lifestyles. Similarly design, education and policy interventions must take heed of the competing discourses and expectations manifest as and through homemaking and must accept that these latter imperatives will always feed back into and onto such interventions, shifting the socio-technical infrastructures, systems and behaviours articulated through and around homemaking. This understanding is growing, with a body of sustainable design research and practice emerging that positions practice as facilitation and negotiation of community expectations, such as the Sustainable Everyday Project (Manzini and Jégou 2003). As I have previously argued, it is vital that sustainability does not require heroism on the part of the householder and that it be engaged with the actual complexities of socio-technical housing infrastructure, expectations and trajectories (Crabtree 2006). Further, our understandings and expectations of sustainability must accommodate messiness and be able and willing to shift over time as tensions and opportunities emerge.

Homes, economies and identities are intricately bound up in processes operating at multiple scales and have always been enmeshed in overt and covert moral systems and debates. Sustainability reveals the deeply political nature of the ostensibly private realm of home, as long proclaimed in postmodern and feminist theory (for example Hayden 1981, Honig 1996, Kaika 2004). The challenge of sustainability is in dealing with the diverse socio-cultural, economic and moral imperatives that are rendered overt when we recognise our homes for the utter conceptual mess that they are, in line with Mouffe's (1995) agonistic model of democracy, and being okay with that.

References

Amin, A. 2010. 'Cities and the Ethic of Care among Strangers', seminar paper, Centre for Research on Social Inclusion and the Department of Environment and Geography, Macquarie University, 16 February 2010.

Crabtree, L. 2006. Disintegrated houses: exploring ecofeminist housing and urban design options. *Antipode*, 38(4), 711–734.

Hayden, D. 1981. *The Grand Domestic Revolution: A History of Feminist Designs for American Homes, Neighbourhoods and Cities*. 2nd Edition. Cambridge, MA: MIT Press.

Honig, B. 1996. Difference, dilemmas, and the politics of home, in *Democracy and Difference: Contesting the Boundaries of the Political*, edited by S. Benhabib. Princeton: Princeton University Press, 257–277.

Kaika, M. 2004. Interrogating the geographies of the familiar: domesticating nature and constructing the autonomy of the modern home. *International Journal of Urban and Regional Research*, 28(2), 265–286.

Manzini, E. and Jégou F. 2003. *Sustainable Everyday, Scenarios of Urban Life*. Milan: Edizione Ambiente.

Mouffe, C. 1995. Feminism, citizenship, and radical democratic politics, in *Social Postmodernism: Beyond Identity Politics*, edited by L. Nicholson and S. Seidman. Cambridge, UK: Cambridge University Press, 315–331.

PART III
Governance and Citizenship

Chapter 8
Mapping Geographies of Reuse in Sheffield and Melbourne

Matt Watson and Ruth Lane

Introduction

'Reuse' sits just below 'reduction' as one of the most favoured options in the 'waste hierarchy' (Gertsakis and Lewis 2003). The hierarchy (Figure 8.1), cited as a guiding principle in international policy statements, is part of a gathering paradigm shift in approaches to governing resources and waste: from a linear model ending in disposal to an increasingly cyclical model which seeks to minimise final waste volumes (Watson, Bulkeley and Hudson 2008). Australia and the United Kingdom share in the progress towards new models of resources management, with recent policy statements in both countries affirming bold intention. However policy initiatives in pursuit of these objectives, at least as they relate to household waste, have focused on recycling (where objects are disaggregated into component materials for the manufacture of new goods) rather than reduction (which means consuming less), or reuse (where objects retain material integrity and are put into use again).[1] Limited policy engagement with reuse can be seen to indicate the complexities involved for governance when it intervenes in the processes and practices through which goods travel from situations in which they do not have enough value to be kept, to those where they have enough value to be newly acquired.

In this chapter we engage with the problematic relations between modes of governing waste and the routes that things must travel – through infrastructure, spaces, practices, values and meanings – to be reused. Towards this end we draw on two key bodies of existing geographical research to focus on how different systems of governance in Melbourne (Australia) and Sheffield (United Kingdom), influence the effective circulation of household goods. In the first of these bodies of literature, a number of authors have focused on the governing of municipal waste management as it is reshaped by the sustainability agenda and related understandings of environmental citizenship (Davoudi 2000, Bulkeley, Watson and Hudson 2007, Watson, Bulkeley and Hudson 2008, Bulkeley and Gregson 2009). From the other side, offering means of accessing the practices around reuse,

---

1   For the purposes of this chapter, we see reuse as involving some exchange, leaving aside the reuse of objects – such as carrier bags, rechargeable batteries, washable nappies or refillable containers – by one owner or household.

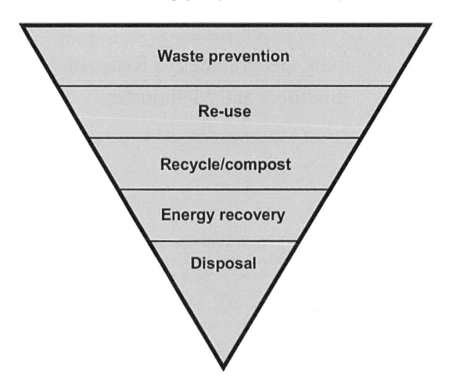

Figure 8.1   Waste Hierarchy
Source: *Waste Strategy for England 2007*, Department for Environment, Food and Rural Affairs

are geographies of retail and consumption, which have proliferated over the last 20 years. The practice of consumption, once depicted as the end point of a commodity chain, is now viewed in more complex ways, seen as emerging from the intersection of an active consumer with systems of production and broader socio-technical processes (Crewe 2000, Crewe, Gregson and Brooks 2003, Gregson, Crewe and Brooks 2002b, Gregson, Metcalfe and Crewe 2007b, Shove 2003, Shove 2006). The study of second-hand household goods has presented opportunities for new understandings of consumption practices in the home as it entails disposal by one owner as well as acquisition by another. It also throws some light on the informal channels that are particularly prominent in the movement of second-hand goods (Williams 2002, Williams and Paddock 2003, Williams and Windebank 2005, Gregson 2007, Lane, Horne and Bicknell 2009).

Harriet Bulkeley and Nicky Gregson (2009) have argued the need to bring these literatures on governance and consumption practices together to advance understandings of whether and how more effective waste reduction can be achieved through governance arrangements. Following this same agenda, focusing particularly on reuse, we begin by setting out the policy and governing framework in relation to reuse in Melbourne and Sheffield. From there we explore some of the more common routes of reuse in the two cities, considering the range of actors (whether government, business or not-for-profit organisations), the various rationalities (economic, environmental or social) invoked in different reuse channels, and some of their spatial and temporal characteristics, along with issues of infrastructure. We move on to gather together existing knowledge and theorisation of the uses and meanings of second-hand goods in both countries. Finally we characterise the disjunctures between waste policy and the routes of reuse, and note their initial implications.

Policy and Governance Frameworks in Melbourne and Sheffield

Melbourne is Australia's second largest city and the fastest growing, with a population now approaching four million (Australian Bureau of Statistics 2009). Its growth since the 1990s is primarily due to immigration, and recent government planning initiatives project a population of five million by 2030 (Department of Planning and Community Development 2008). Sheffield is England's fourth largest city but is very small compared to Melbourne, with a population of just over half a million. By 2020 its population is projected to increase by 12 per cent on 2001 levels (Office of National Statistics 2008).

Both cities are obliged to pursue ambitious targets relating to waste management. Key commitments in Melbourne follow from *Sustainability in Action: Towards Zero Waste Strategy* (Sustainability Victoria 2005), which provides a ten year plan for waste management in the state of Victoria, setting out targets, including one for municipal waste of 65 per cent recovery (by weight), through reuse, recycling or energy generation, by 2014. Sheffield is subject to UK national targets set under the *Waste Strategy for England 2007* (DEFRA 2007), which sets forward a new target to reduce the amount of waste not being reused, recycled or composted by 29 per cent (by weight) of the levels in 2000, by 2010, ramping up to 45 per cent by 2020. Both policy documents repeat their respective states' commitment to the waste hierarchy as a guiding principle. Resulting from the inheritance of policy framing, institutional configurations and competence and infrastructures from the era of disposal, recycling has proven more immediately amenable to policy intervention than either reuse or reduction (Watson, Bulkeley and Hudson 2008, Bulkeley and Gregson 2009). However in the *Waste Strategy for England 2007*, the first in a list of key objectives includes the need to 'put more emphasis on waste prevention and reuse' (DEFRA 2007: 3). Victoria's *Sustainability in Action: Towards Zero Waste Strategy* bundles reuse along with waste reduction in statements about targets for the various waste streams (Sustainability Victoria 2005: 11) and the new *Metropolitan Waste and Resource Recovery Plan* for Melbourne acknowledges a role for 'demonstration projects for re-use and recovery, e.g. collection of second-hand items using existing recycling bins' (Department of Sustainability and Environment 2009:8).

Waste management policy and associated reduction targets in the United Kingdom are strongly influenced by top down imperatives set out in the EU Landfill Directive. It appears that Victorian waste management policy makers have more autonomy, which potentially allows more leeway for experimentation with new initiatives for waste diversion. A key incentive for meeting reduction targets in the United Kingdom is the application of a significant landfill levy, now at £40/tonne (A$77/tonne), with scheduled annual increases through to 2013. By contrast the landfill levy in Victoria is only A$9/tonne, suggesting that policy initiatives rather than price signals are expected to drive the reduction. Melbourne's large size and diverse industry base may make it easier for new initiatives around reuse to develop viable economies of scale than in Sheffield.

The governing of municipal waste management is not something restricted to the formal institutions of government. Over recent decades in both the United Kingdom and Australia the range of actors involved has steadily grown, reflecting two distinct dynamics. Firstly, a general shift of governing responsibility from the nation-state to other tiers of government, but also to a range of non-governmental institutions, from multinational corporations to local community groups, a trend captured under some meanings of the term 'governance'. In waste management this is reflected most clearly by the shift of operational responsibility from local authorities to commercial contractors. Secondly, the shifting waste agenda has meant that operational waste management is increasingly exceeding the capacity of government bodies. In comparison to moving waste from one bin per household to a centralised disposal facility (a landfill or incinerator), imperatives to recycle and to reduce and reuse waste demand intervention and participation in complex relations and processes – from household habits and routines to international materials markets.

This shift, from a linear to an increasingly cyclical understanding of the flow of materials, has entailed major challenges throughout scales of governing. Strategy documents and radically improving figures for recycling both indicate that this shift is increasingly well established. However the challenges of recycling appear simple when compared to the complex and often micro-scale relations involved in routes of reuse. In the *Waste Strategy for England 2007* Figure E1 (Figure 8.2) represents the dominant logic of material flow in a paradigm dominated by recycling, in which there is no hint of the smaller circularities entailed in reuse, such as between consumers and between consumers and retailers.

Routes of Reuse in Melbourne and Sheffield

Infrastructure, Processes and Practices

Routes of reuse are deeply diverse, each providing different means for 'stuff' to overcome time and/or space to get from situations of divestment to situations of acquisition. In briefly trying to sketch this diversity, it is possible to construct a

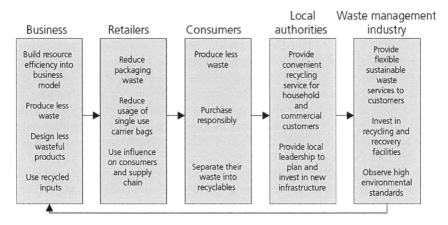

| | | Local | Waste management |
Business	Retailers	Consumers	authorities	industry
Build resource efficiency into business model	Reduce packaging waste	Produce less waste	Provide convenient recycling service for household and commercial customers	Provide flexible sustainable waste services to customers
Produce less waste	Reduce usage of single use carrier bags	Purchase responsibly		Invest in recycling and recovery facilities
Design less wasteful products	Use influence on consumers and supply chain		Provide local leadership to plan and invest in new infrastructure	
Use recycled inputs		Separate their waste into recyclables		Observe high environmental standards

Figure 8.2   Extract from Figure E1 in *Waste Strategy for England 2007*

continuum based on the extent of mediation between the person getting rid of a thing and another person acquiring it. Placing locales and forms of transaction on this continuum is not a precise science, as there is a range of forms of mediation, but the continuum roughly runs from commercial second-hand shops through to direct gift relations between friends, relatives and neighbours. What follows is by no means comprehensive, but seeks to pick up major routes of reuse and in so doing show their overall diversity.

High street second-hand retail outlets are perhaps the most visible locale in routes of reuse in both cities. In Sheffield the majority of these are charity shops, which typically receive donations and sell what is worth selling. In Melbourne, in addition to charity shops, there is a plethora of commercial second-hand shops, notably retro and vintage clothes retailers in trendy city quarters but also a wide range of retailers specialising in specific products such as whitegoods, furniture or musical instruments. These can represent still more mediated routes of reuse, acquiring goods from second-hand supply markets that have an international reach. Second-hand retailers have been the focus of some consumption-oriented social science research in the United Kingdom, which demonstrates the complex range of motives and values that converge in such shops and the goods they sell (Goodall 2000, Horne 2000, Gregson, Crewe and Brooks 2002a).

Second-hand retail can be a feature of some local social enterprises. In both Melbourne and Sheffield there has been a proliferation over the last ten years of new enterprises that involve a mix of government, business and charity organisations for the purpose of delivering goods or services with social benefits. These social benefits may include the provision of repaired domestic goods to low-income clients and various forms of skills training. For example in Melbourne, a charity organisation primarily involved in providing employment training and placements for people with disabilities is now assuming management of a number of waste transfer stations where recyclable materials are sorted from general waste and reusable objects are salvaged for sale in a tip shop (Table 8.1). These multi-objective social enterprises frequently are conduits of reused goods which do not undergo repair, notably items of household furniture. 'Renew', an innovative project that has recently been trialled in Melbourne (Table 8.1), involves the collection of reusable goods from kerbside recycling bins the day after the standard collection for recyclable waste (Agostino 2008). Sorting and repair or remanufacturing is then conducted through a charity organisation supporting homeless youth. This was initiated by a commercial business and involves partnership agreements with state and local government agencies and charity organisations, as well as in-kind support from a range of other businesses.

A significant body of existing research explores relatively informal spaces of exchange in the United Kingdom, notably car boot sales (Gregson and Crewe, 1997b, Gregson and Crewe 1997a, Williams 2002, Williams and Paddock 2003) and second-hand fairs (Chantelat and Vignal 2002). These also exist in Melbourne, alongside 'garage sales' and school and church fetes. While prices may be set initially, they are often negotiated through bartering. As well as an opportunity for acquiring goods for free, these venues also draw visitors for their interest, entertainment or opportunities for sociality either at the neighbourhood level or in the community of people who frequent these events as sellers and buyers.

In contrast to shops, car boot sales and garage sales (with garages providing spaces for things to be gathered over time and for interacting with customers), there are a range of mediated means for people with something they no longer want to find someone who may want to acquire it. Local newspapers and classified freesheets have played this role for a long time. Increasingly this role of mediation is performed by the Internet, most obviously by eBay™. The Internet has enabled sellers to reach specialist markets over wide geographic areas and the means to have buyers bid against each other, meaning goods can often have both a better chance of selling and of reaching a higher price. However the Internet has also enabled a range of other second-hand exchange relations, from traditional local classified advertisements going online to special interest bulletin boards enabling specialist equipment to find second-hand markets, through to free giveaway sites such as local networks under the international Freecycle™ concept.

Existing provision for waste disposal provides a number of opportunities for reuse. These run from formalised schemes, notably tip shops where salvaged goods are available for sale, to informal practices that piggy-back on existing waste disposal services. In Melbourne, and in some other Australian cities, a service is provided to residents by local government for the collection of more bulky items that are placed on the kerbside in front of the residence in a given period of time. While intended as a waste collection and disposal service with a measure of materials recycling, hard rubbish collections arouse much interest among residents as they provide opportunities for acquiring goods as well as disposing of them, all free of charge. From the moment items are placed out on the kerbside strip they are subject to scavenging by both local residents and professional or semi-professional scavengers and collectors. The 2007 survey of Melbourne households found that over 40 per cent of respondents had acquired things by scavenging hard rubbish collections in the previous two years (Lane, Horne and Bicknell 2009).There is no similar regular opportunity for scavenging in Sheffield, with scavenging limited to opportunistic appropriation from skips on the street and occasional kerbside dumped rubbish.

Finally, passing on of goods within personal social networks is a major but largely hidden and unmeasurable means of exchange of second-hand goods. Such exchanges typically occur in relationships unmediated by any institution, commercial organisation or exchange of money (but which potentially carry heavily laden personal social obligations and complex expectations of reciprocity). These informal and unmediated routes of exchange have attracted some interest, again revealing insights into the role of household goods in people's everyday lives and relationships that are obscure to analyses that see reuse as overwhelmingly a market phenomenon defined by purchase decisions (Miller 1998, Gregson and Beale 2004). In the 2007 survey of Melbourne householders, 74 per cent of respondents reported giving goods away to family or friends in the last two years, with 62 per cent acquiring goods through such channels. There is evidence that these channels are used more extensively by those who have strong social connections in the region, and that they are less likely to be used by recent immigrants (Bulkeley and Gregson 2009, Lane, Horne and Bicknell 2009). These channels are especially likely to be used by households with children, reflecting both a greater need to acquire and dispose of goods than other households and perhaps also an association between there being children in households and those households having more extensive social networks (Lane, Horne and Bicknell 2009).

Table 8.1 provides a schematic comparison of a more comprehensive listing of routes of reuse. It highlights the multiple actors, locales, processes, rationalities and motivations involved. Of particular note is the frequency with which multiple actors and multiple rationalities underpin some of the more mediated routes of exchange, such as waste transfer stations or second-hand retail shops. In many of the channels outlined in Table 8.1 formal (that is market) and informal (that is gifting or scavenging) exchanges coexist.

Table 8.1   Common Routes of Reuse in Melbourne and Sheffield

	Agents and Agencies	Rationalities Involved	Relationships and Transactions	Charges	Technologies and Infrastructure	Locations	Temporality	Material Values
Second-hand retail shops	Commercial businesses (shops involving charities are included under the social enterprise model above)	*Multiple* – mainly economic, but including curiosity, nostalgia and collectors interests	*Formal* – subject to standard government regulations for retail premises and businesses	Clients visiting shop purchase items either for set price or by bartering. Businesses pay rent on retail premises	*Infrastructure* – retail premises *Transportation* – vehicles to transport goods from sources such as deceased estate auctions to the retail premises *Communications* – advertising signs	Shops usually in High street locations	Shops open normal retail trading hours throughout the year	Objects in their original form valued as reusable goods or collectables with potential value as antiques
Online retail sites and classified newspaper adverts	Large and sometimes multinational companies such as eBay™, Gumtree, SENSIS and incorporated newspaper classifieds such as Domain.com or *The Trading Post.*	Profit, convenience	*Formal* – subject to government regulations for retail practices but difficult for governments to police	Individuals sell items to other individuals for either a set or negotiated price. Company owning the site may charge a transaction fee	*Transportation* – postal services, courier services, domestic vehicles, *Communications* – web sites	Objects move directly from the seller to the purchaser, usually by post or commercial delivery vehicle	All hours all year round	Objects in their original form. Residents as buyers and sellers

More Mediated ←

	Agents and Agencies	Rationalities Involved	Relationships and Transactions	Charges	Technologies and Infrastructure	Locations	Temporality	Material Values
Social enterprises e.g. Renew project in Melbourne	A combination of business, government, community or charity organisations	*Multiple* – social and economic benefits, eco-efficiency	*Formal* – partnership agreements between government, business and charity organisations that usually involve some contractual agreements, employment contracts and training programs *Informal* – donations made by individual householders, volunteering of labour and resources, donations of various forms of 'in kind' support	Materials usually donated free of charge Aim to develop new business opportunities around resale and remanufacturing that generate sufficient income returns to fund the program over the longer term May be initially subsidised through cash and in kind support from various government and industry partners in the project	*Infrastructure* – kerbside recycling bins, charity bins, collection bags, shops etc. *Transportation* – collection vehicles, warehouse storage facilities and sorting facilities *Communications* – shop signs, leaflets in mail boxes, web sites etc.	kerbside, shopping centres, warehouses in industrial areas	Varied – some collection facilities such as charity bins are available all the time, door-to-door collections may occur once or twice a year Acquisitions made through retail outlets may occur year round.	Individual objects valued for resale, reuse or remanufacturing potential

	Agents and Agencies	Rationalities Involved	Relationships and Transactions	Charges	Technologies and Infrastructure	Locations	Temporality	Material Values
Melbourne Waste transfer stations e.g. Darebin Resource Recovery Centre	Local governments, Sustainability Victoria, charity organisations, commercial waste contractors	*Multiple* – reducing waste to landfill, reducing environmental impacts, recovering value and social and economic benefits for employees	*Formal* – councils own the facilities and contract either waste management businesses or charity organisations to run them with financial support from State-govt agency Sustainability Victoria for enhancing recovery of recyclable materials Commercial transactions involved in charges for householders to use the facility and for purchases from tip shop	Fee charged to individuals as they arrive at the gatehouse based on assessment of load Fee charged by landfill operators to Darebin Resource Recovery Centre based on weight of materials disposed of in landfill Fee charged for recovered objects purchased in tip shop	*Infrastructure* – waste transfer facility with sorting equipment and retail space *Transportation* – vehicles for moving materials around the site and from the site to the landfill Vehicles used by individuals to transport materials from residences to the facility *Communications* – web site, explanatory signs at gatehouse and within the facility	5 sites in suburban areas of Melbourne	Facility open 7 days a week throughout the year	*Multiple* – discarded objects and materials classified separately into material type and objects with resale potential

	Agents and Agencies	Rationalities Involved	Relationships and Transactions	Charges	Technologies and Infrastructure	Locations	Temporality	Material Values
Sheffield Household waste recycling centres	Contractors (Veolia), local government	Waste management to meet environmental standards and optimise materials recovery	*Formal* – contractual relation between council and contractors *Informal* – limited opportunities for informal scavenging	No fees for householders unless in a commercial vehicle without a license	*Infrastructure* – separate skips and designated areas taking a wide range of separated materials for recycling, composting, safe disposal (batteries, engine oil), energy recovery and landfill	5 sites in suburban areas of Sheffield, some on old or current landfill sites	Facility open 7 days a week throughout the year	*Multiple* – discarded objects and materials classified separately into material type for material or energy recovery
'Alternative' sites of exchange (i.e. car boot sales, fairs, gararge sales etc.)	Voluntary organisers who may be individuals or organisations such as schools or churches	Economic, recreation or other social value	*Semi-formal* – managed according to established rules by voluntary organisers	Sellers set prices and buyers negotiate by bargaining	*Infrastructure* – temporary stalls erected *Transportation* – objects usually exchanged face-to-face and transported by private vehicle. *Communications* – promotional signage, public media	Public parks or showgrounds or the grounds of schools or churches Objects move from private seller to stall to buyer		Objects in their original form valued by sellers for economic return and by buyers for reuse potential, collectors interest or curiosity

	Agents and Agencies	Rationalities Involved	Relationships and Transactions	Charges	Technologies and Infrastructure	Locations	Temporality	Material Values
Charity bins and collections	Charity organisations, sometimes contractors	Supporting the needy	*Formal* – subject to government regulations for incorporated not-for-profit organisations May engage contractors to clear bins and process materials *Informal* – donations made by householders	No charges to householders – a means of collecting goods that may then be sold for a profit to fund programs for the needy	*Infrastructure* – dedicated bins, collection bags, sorting facilities *Transportation* - vehicles to empty bins *Communications* – notices in mail, public media promotions	Bins located on suburban streets, sorting usually conducted in warehouse facilities	Varied – charity bins available all hours year round but door-to-door collections may occur once or twice a year	Objects in their original form Residents as donors
Online exchange/give away sites	International network of voluntary moderators	Convenience, supporting the needy, anti-consumerism	*Semi-formal* – managed by voluntary moderators according to agreed rules that forbid commercial transactions	Strictly no fees involved	*Transportation* – postal services, courier services, domestic vehicles *Communications* – web sites	Objects move directly from donor to acquirer, usually by some form of personal transport	All hours all year round	Objects in their original form valued for reuse potential

➤ More Direct

	Agents and Agencies	Rationalities Involved	Relationships and Transactions	Charges	Technologies and Infrastructure	Locations	Temporality	Material Values
Scavenging of Melbourne's hard rubbish collections	local government, state govt (via model contracts), waste management contractors, individual scavengers	*Multiple* – local government motivated to satisfy public demand, and prevent illegal dumping (considered a form of environmental pollution) Individual scavengers motivated by economic opportunities to acquire useful items free or sell them for profit	*Formal* – contractual arrangements between local government and contractors *Informal/illegal* – activities of individuals piggy-back on local government waste management system Scavenging considered an illegal transaction by some councils	Service funded via council rates, councils pay contractors to implement, contractors make some additional income from selling on scrap metal, scavengers may also make income from sales of scavenged goods	*Infrastructure* – nil *Transportation* – scavengers' vehicles, contractor's trucks. *Communications* – public notices and mailed information to householders	Kerbside, residential dwellings, scrap metal yards	Annual or at call collections	*Multiple* – contractors and local governments treat municipal waste as lumped category with negative value Scavengers either value specific objects for potential use or resale, or component materials for market value as scrap

	Agents and Agencies	Rationalities Involved	Relationships and Transactions	Charges	Technologies and Infrastructure	Locations	Temporality	Material Values
Gifting among family and friends	Networks of friends and relatives	Social obligations and reciprocal relationships, supporting the needy	*Informal* – no formal institutional relationships involved	Usually no payments involved, although may be elements of exchange	*Transportation* – private vehicles or postal/courier services	Objects move directly from one domestic household to another, usually by some form of personal transport	All year round although there may be peak periods of activity for specific households associated with major changes such as moving house or death of a family member	Objects in their original form valued for their social associations as well as their reuse potential

The Uses and Meanings of 'Second-hand' in Melbourne and Sheffield

A review of existing literature and our own research shows that the range of purposes, meanings and motivations around second-hand goods is at least as diverse and complex as the infrastructure, spaces, temporalities and rationalities involved in their routes of reuse.

While the key driver for the political prioritisation of reuse is environmental concern, evidence for how far concerns over environmental impact are significant in attitudes towards reuse is very varied. In short, it appears that if people are asked about the environment they will voice strong concern. If, however, they are asked about why they buy reused goods, the environment will figure considerably less prominently, if at all. Ethics also translates to the directly social, particularly in terms of decisions to donate goods to worthy causes, which can be framed as altruistic. This is evidenced by survey findings in both the United Kingdom and in Melbourne that show respondents are much more likely to give things away either to charities, or through their own social networks, than they are to sell them (Association of Charity Shops 2006, Lane, Horne and Bicknell 2009).

Aside from what might be expected in terms of ethics around reuse, the fact that reused goods are relatively cheap is the dominant issue in mainstream accounts of why people buy them. Clearly, so far as reused and remanufactured goods are cheaper than new ones, there is likely to be alignment between buying reused goods and being relatively poor. The continuing validity of this relationship is borne out by a number of studies with relatively deprived consumers (Williams 2002).

However the picture of variation in attitudes to reused products according to social status is far more complex than this. Firstly, declining real-term costs of many new commodities, notably clothes and home appliances, has reduced the price advantage of reused goods and brought new goods within reach of the relatively deprived (McCollough 2007). Secondly, a range of quantitative studies has shown that the majority of users of alternative retail channels are not necessarily to be considered socially marginalised (Williams 2002), with affluent social groups making up a substantial proportion of purchasers at both car boot sales (Stone, Horne and Hibbert 1996) and charity shops (Mintel 1997, Mintel 2000, Williams 2002).

The use of second-hand outlets through choice rather than necessity opens up the range of different meanings and valuations that various second-hand goods can have. Interviews with purchasers at car boot sales show the many reasons that people can have for purchasing second-hand beyond financial necessity, highlighting how second-hand exchange can be about fun, sociality and the considered pursuit of distinctive style (Gregson and Crewe 2003, Gregson and Crewe 1997a). Nicky Gregson and Louise Crewe (2003) argue that, for many car boot sale purchasers, second-hand goods are appropriated for reasons parallel to the purchase of high status new goods, such as the pursuit of distinctiveness, uniqueness and individuality. This clearly carries through into motivations for the purchase of antiques, where the signs of previous ownership and use can become part of what is valued. At the same time second-hand goods can be part of the construction of a deliberately anti-consumerist, anti-corporate or otherwise ethically driven consumption identity. However based on 120 face-to-face interviews in socio-economically distinct areas of Leicester, Colin Williams (2002, Williams and Paddock 2003) argues that this reassessment of second-hand exchange has validity only in relation to the relatively affluent. Interviews with the relatively deprived found that economic necessity remained the main motivation for using informal and second-hand means of acquisition. The ability to recognise the positive potential of second-hand goods is itself not independent of social status. While not dependent on financial wealth, it can take a certain amount of cultural capital to be prepared to engage creatively with second-hand retail environments.

Ethics, economics and the symbolic potential of second-hand therefore all play a role. However recent research has increasingly emphasised less abstract themes, recognising the power of stuff itself to impel owners to seek means of discarding it, whether for reuse or not. Research on divestment practices has highlighted both temporal and spatial dimensions that together contribute to a certain 'lumpiness' in material flows across the life stages and events within a household unit. Nicky Gregson and her co-researchers (Gregson, Metcalfe and Crewe 2009, Gregson 2007, Gregson, Metcalfe and Crewe 2007a) highlight events such as a death in the family, moving house or major renovations as being key periods when an excess of household goods must be dealt with. At such times the ease and convenience of channels for disposal is likely to be important. Having space at home where objects can be stored or sorted is a factor in the decision to keep or discard used or damaged goods (Bulkeley and Gregson 2009). The Melbourne survey found that the best overall predictor of the likelihood of a household both acquiring and disposing of used goods was the presence of children and of a parent, usually a mother, at home (Lane, Horne and Bicknell 2009).

Finally stuff makes itself felt also through the responsibilities people can experience for things. This material responsibility is a theme generally submerged by dominant accounts of a consumerist and materialistic society, but brought to light by authors in a number of ways. Ethics of care towards possessions and the impacts they entail are more complex and nuanced than is captured by environmental concern (Gregson, Metcalfe and Crewe 2009). Nicky Gregson and Louise Crewe (2003) argue that passing goods on to a further use, whether through sale or donation, is partly about the responsibility people feel to durable possessions in which they recognise persistent embedded value. They suggest that a conservative ethics of care was a significant part of respondents' accounts of why they participated in second-hand exchange, an ethics with only tenuous connections to the environmental or social implications of buying things new. Similarly Tim Cooper (2005) found a sense of responsibility to possessions, with respondents in his study commonly reporting the desire that items to be disposed of go to some further good use. Where objects stand for significant people – such as the particular possessions of deceased relatives – a sense of responsibility for things becomes more complex than a sense of responsibility simply for the material.

In summary, several patterns emerge in the motivations to use second-hand channels to dispose of or acquire goods. In disposal, householders are clearly influenced by the convenience of particular channels, particularly if they need to dispose of a large number of items in a short space of time or if they have a frequent need to dispose of things, as might be the case in a household with young children. They are also motivated by a desire to help people in need, including friends and relatives. While householders may have a general desire to avoid waste, they may also feel a sense of stewardship towards a specific object and feelings of responsibility for its integrity and ongoing use or value. Acquisition of used goods is strongly influenced by economic necessity; however this may vary over the life course with a greater need to acquire things during child-raising years, or when moving house. The presence of social networks among family and friends strongly influences access to informal gift-giving arrangements. However curiosity and collectors' interests are also motivations for the acquisition of used goods, particularly through charity shops or alternative retail sites.

Disjunctures of Policy and the Routes of Reuse

The brief discussion in the last two sections begins to draw out the disjunctures between the rationalities of policy around waste management and minimisation, and the complexities of actors, processes, spaces, meanings and purposes involved in routes of reuse. There are clear disjunctures between the singular rationalities of waste management policy and the multiple rationalities of effective reuse channels. It is clear that in only a minority of situations of reuse is a concern for environmental sustainability a primary motivation, and often it does not figure at all. Things are pushed and pulled along the routes of their reuse by complexes of purpose and motivation, ranging from relatively simple economic interest to explicit anti-consumerism; from desire for a uniquely 'new' object to the need for a specific functional item at the lowest possible cost; from the compulsion to be rid of things to a sense of responsibility to the object or the history it represents. Even in Freecycle™, which is presented as entirely motivated by keeping stuff out of landfill,[2] and in which there is no expectation of any form of direct reciprocity between giver and receiver, participants show a wide range of other motivations in offering as well as requesting goods. Yet policy commitments to promote reuse are framed overwhelmingly in discourses of environmental responsibility, legitimated with reference to the waste hierarchy.

To what extent should governance initiatives recognise these channels or seek to support them? To engage with informal channels of reuse would require

---

2   'Our mission is to build a worldwide gifting movement that reduces waste, saves precious resources & eases the burden on our landfills while enabling our members to benefit from the strength of a larger community.' www.freecycle.org/about/missionstatement [accessed 14 February 2011].

a much greater level of coordination between waste management agendas and social welfare agendas. There are potential tensions between these agendas and the economic agendas of commercial contractors. While it may be possible to balance these in small-scale social enterprises, this might become more difficult to manage if they were to be 'scaled up'.

There are also practical challenges in scaling up the reuse of household goods as opposed to materials recycling. High levels of recycling require engagement of waste management strategies with household practices (Darier 1996) but the challenges of getting materials from individual households to conduits of materials recycling can be seen to be continuous in many ways with the infrastructure and competences needed for moving waste from homes to landfill sites (Watson, Bulkeley and Hudson 2008). In comparison, to be reused as physically coherent objects, things follow much more diverse and specific routes. Reuse in both cities examined in this chapter comprises a bewildering array of actors, means of travel of objects, and storage and exchange spaces, as evidenced in Table 8.1. Further, as most reuse happens to materials that do not enter the waste stream in the first place, the state has relatively little control over these routes. At a basic pragmatic level, then, it is difficult for the state to intervene to promote reuse in any general sense. This is reflected in the concentration of policy initiatives on forms of reuse which do not involve exchange between successive uses, such as washable nappies and reusable carrier bags (for example Sheffield City Council 2009).

So at the level of discourse there is a disjuncture to be overcome, between the singular governing rationale behind council waste strategies and the diversity of motives, meanings and purposes which can power the processes of reuse – processes which serve those strategies' targets. But discourses do not exist independently of other aspects of the social. They emerge from and are reproduced by institutional structures, in frameworks of law and policy and in the competencies and practices of policy makers, council officers and workers. For example in the United Kingdom the EU Landfill Directive drives targets and policy for waste management in a top-down manner which then has the effect of establishing divisions of responsibility within government that make it difficult for councils to work effectively with multi-objective social enterprises. Offering homeless people accommodation and work does not help fulfil any of the targets driving UK local authorities' waste management. By comparison, the mix of social and environmental rationales at play in Melbourne's waste transfer stations may be more possible where government agencies have more autonomy to develop their own policy agendas. However a mismatch between structures of governing and some routes of reuse can impose direct limitations on some reuse channels, such as the criminalisation of reuse via the scavenging of hard rubbish collections in Melbourne.

Finally, it is undeniable that many reuse channels do not sit easily with dominant neo-liberal ideologies of public policy. To put it baldly, advancing waste minimisation, including through reuse, means people buying less stuff. Ultimately both reduction and reuse cannot be pursued far before they conflict with the conventional understandings of economic growth that dominate local, regional and national development strategies.

There are, then, profound policy challenges to promoting reuse in any general way. Indeed to the extent that reuse occurs through informal and unregulated channels of more or less direct giving and receiving, there are arguments for keeping the state's involvement within limits to prevent these channels being unnecessarily impeded. Nevertheless from our explorations so far, it is clear that there are a number of points of intervention which could usefully be explored to inform any conscious effort to scale up reuse channels.

Firstly, recognition of the potential synergies between social objectives – whether raising funds or providing skill development and work for disadvantaged people – and sustainable resource management could enable the development of major, and highly visible routes of reuse. There are some examples of this in Sheffield, but nothing on the scale of Melbourne's Resource Recovery Centres, where a third sector organisation manages the flow of materials into the stations to maximise recovery and reuse while simultaneously serving multiple social objectives.

Secondly, authorities and contractors should be alert to possibilities for making slight changes to existing infrastructure to enable more or less formal routes of reuse. For example waste transfer stations could more often have designated areas and procedures for capturing objects with significant reuse value. Provisions to enable reuse range from considered physical space for stuff to wait for the moment of acquisition through to the considered use of variable charges and incentives for both giving and acquiring. Melbourne's waste transfer stations, while charging for disposal, also offer opportunities for cheap purchases.

Thirdly, opportunities could be explored for enabling reuse at those life moments when individuals and households tend to be overcome by demands to acquire or be rid of stuff – such as when moving house, downsizing, co-habiting the first time or dealing with the death of a family member. At times of high need, the most convenient channels are likely to be used, that is, those available year round at convenient times of day.

What Do We Need to Know for Policy to Better Promote Reuse?

Recognising that existing routes of reuse form a complex system, it is clear that designs for initiatives to advance reuse need to be alert to potentially negative consequences. For example it is important to understand any potentially negative consequences of promoting formal schemes involving contractual arrangements and market transactions for informal transactions involving gifts, donations, free acquisitions and informal types of reciprocity. Relatedly, more work is needed to understand the tensions and potential synergies that arise from the interplay of different forms of social, environmental and commercial agendas.

A crucial theme here is that of scale. A common criticism from proponents of ecological modernisation is that reuse is better considered a 'cottage industry' rather than a significant means for reducing consumption or waste. If reuse is to be promoted there need to be far more local, small-scale reuse routes or else routes on a larger and therefore more formal scale. Does scaling up reuse have implications for moving away from networks of individuals operating with a social benefits motivation towards more formal business arrangements with a profit motivation? When recycling initiatives began in the 1970s they were largely driven by environmentally conscious individuals and community organisations. However as they were scaled up and supported through government subsidies new large-scale businesses emerged, such that materials recycling is now the domain of multinational companies (Koponen 2002). Is this a desirable model for a scaled up reuse industry?

A final concern is the extent to which reuse may play an ambivalent role in reducing overall resource consumption. Increasing possibilities for original owners to realise value – or at least to obviate any guilt – in the process of divestment could risk increasing the 'churn' rate of consumption for those concerned to buy new. At the same time the growing availability of second-hand goods enables people on a given income to buy more things, or higher status things. It is possible that to some extent increasing reuse will simply increase the number of transactions particular objects go through without reducing the overall throughput of resources. Alternatively, reuse could function as a strategy for moving towards lower consumption levels by extending the lifespan of common household goods and promoting an ethic of care or stewardship for goods more generally.

While there are clearly many aspects of reuse that warrant further research, two key areas stand out. First, further study of household practices of reuse is needed that considers how these link up with the physical infrastructure of the home and neighbourhood. For example, by contrasting inner suburbs with high population density and smaller dwellings with the new suburban estates being developed on the urban fringe which tend to be lower density with larger dwellings. A significant concern here is the convenience of different channels of discarding in specific socio-demographic and geographic contexts. Second, case study research on social enterprises is needed. Given that these are where the more innovative initiatives for reuse are coming from, they warrant closer study, involving monitoring of their development from the start-up phase to a stand-alone business model. Such case studies would advance understanding of the tensions entailed in balancing multiple objectives and help identify opportunities for greater synergies between policy and practice in different areas of government. In particular, they may offer insights into the issues associated with scaling up initiatives arising from charity, business or government sectors. Research, including ours, has so far only scratched the surface of the complex, problematic relations between waste minimisation policy and the routes of reuse. Pushing understanding of this field further falls within a gathering agenda to understand how governing can better connect with the practices and motivations of householders (Bulkeley and Gregson 2009, Bulkeley, Watson and Hudson 2007).

References

Agostino, J. 2008. *Moonee Valley Renew Trial.* Melbourne: Sustainability Victoria.

Association of Charity Shops. 2006. *An Analysis Into Public Perception and Current Reuse Behaviour Conducted in the East of England. Focusing on Public Attitudes and Perceptions of Reuse through Charity Shops and Furniture Reuse Projects.* London: Association of Charity Shops.

Australian Bureau of Statistics. 2009. *Regional Population Growth, Australia, 2007–08. ABS Cat. No. 3218.0.* Canberra: Australian Bureau of Statistics .

Bulkeley, H. and Gregson, N. 2009. Crossing the threshold: municipal waste policy and household waste generation. *Environment and Planning A*, 41(4), 929–945.

Bulkeley, H., Watson, M. and Hudson, R. 2007. Modes of governing municipal waste. *Environment and Planning A*, 39(11), 2733–2753.

Chantelat, P. and Vignal, B. 2002. 'Intermediation' in used goods markets: transactions, confidence and social interactions. *Sociologie Du Travail*, 44(3), 315–336.

Cooper, T. 2005. Slower consumption: reflections on product life spans and the 'throwaway society'. *Journal of Industrial Ecology*, 9(1–2), 51–67.

Crewe, L. 2000. Progress reports, geographies of retailing and consumption. *Progress in Human Geography*, 24(2), 275–290.

Crewe, L.J., Gregson, N.A. and Brooks, K. 2003. The discursivities of difference: retro retailers and the ambiguities of 'the alternative'. *Journal of Consumer Culture*, 3(1), 61–82.

Darier, E. 1996. The politics and power effects of garbage recycling in Halifax, Canada. *Local Environment*, 1(1), 63–86.

Davoudi, S. 2000. Planning for waste management: changing discourses and institutional relationships. *Progress in Planning*, 53(3), 165–216.

DEFRA, *see* Department for Environment, Food and Rural Affairs

Department for Environment, Food and Rural Affairs. 2007. *Waste Strategy for England 2007.* London: Department for Environment, Food and Rural Affairs.

Department of Planning and Community Development. 2008. *Melbourne 2030: a planning update – Melbourne @ 5 million.* Melbourne: Department of Planning and Community Development.

Department of Sustainability and Environment. 2009. *Metropolitan Waste and Resource Recovery Plan.* Melbourne: Department of Sustainability and Environment.

Gertsakis, J. and Lewis, H., 2003. Sustainability and the Waste Management Hierarchy: A discussion paper on the waste management hierarchy and its relationship to sustainability. Melbourne: Discussion Papers, EcoRecycle Victoria.

Goodall, R. 2000. Charity shops in sectoral contexts: the view from the boardroom. *International Journal of Nonprofit and Voluntary Sector Marketing*, 5, Part 2, 105–112.

Gregson, N. 2007. *Living with Things: Ridding, Accommodation, Dwelling.* Wantage: Sean Kingston Publishing.

Gregson, N. and Beale, V. 2004. Wardrobe matter: the sorting, displacement and circulation of women's clothing. *Geoforum*, 35(6), 689–700.

Gregson, N. and Crewe, L. 1997a. The bargain, the knowledge, and the spectacle: making sense of consumption in the space of the car-boot sale. *Environment and Planning D: Society and Space*, 15(1), 87–112.

Gregson, N. and Crewe, L. 1997b. Performance and possession – rethinking the act of purchase in the light of the car boot sale. *Journal of Material Culture*, 2(2), 241–263.

Gregson, N. and Crewe, L. 2003. *Second-Hand Cultures*. Oxford: Berg.

Gregson, N., Crewe, L. and Brooks, K. 2002a. Discourse, displacement, and retail practice: some pointers from the charity retail project. *Environment and Planning A*, 34(9), 1661–1683.

Gregson, N., Crewe, L. and Brooks, K. 2002b. Shopping, space, and practice. *Environment and Planning D: Society and Space*, 20(5), 597–617.

Gregson, N., Metcalfe, A. and Crewe, L. 2007a. Identity, mobility, and the throwaway society. *Environment and Planning D: Society and Space*, 25(4), 682–700.

Gregson, N., Metcalfe, A. and Crewe, L. 2007b. Moving things along: the conduits and practices of divestment in consumption. *Transactions of the Institute of British Geographers*, 32(2), 187–200.

Gregson, N., Metcalfe, A. and Crewe, L. 2009. Practices of Object Maintenance and Repair: how consumers attend to consumer objects within the home. *Journal of Consumer Culture*, 9(2), 248–272.

Horne, S. 2000. The charity shop: purpose and change. *International Journal of Nonprofit and Voluntary Sector Marketing*, 5(2), 113–124.

Koponen, T.M. 2002. Commodities in action: measuring embeddedness and imposing values. *The Sociological Review*, 50(4), 543–569.

Lane, R., Horne, R. and Bicknell, J. 2009. Routes of reuse of second-hand goods in Melbourne households. *Australian Geographer*, 40(2), 151–168.

McCollough, J. 2007. The effect of income growth on the mix of purchases between disposable goods and reusable goods. *International Journal of Consumer Studies*, 31(3), 213–219.

Miller, D. 1998. *A Theory of Shopping*. Cambridge: Polity.

Mintel. 1997. *Charity Shop Retailing*. London: Mintel.

Mintel. 2000. *Survival of the High Street*. London: Mintel.

Office of National Statistics. 2008. *Subnational Population Projections for England*. London: Office of National Statistics.

Sheffield City Council. 2009. *Draft Waste Strategy*. Sheffield: Sheffield City Council.

Shove, E. 2003. *Comfort, Cleanliness and Convenience: The Social Organisation of Normality.* Oxford: Berg.

Shove, E. 2006. Efficiency and consumption: technology and practice, in *The Earthscan Reader in Sustainable Consumption*, edited by T. Jackson. London: Earthscan, 293–304.

Stone, J., Horne, S. and Hibbert, S. 1996. Car boot sales: a study of shopping motives in an alternative retail format. *International Journal of Retail and Distribution Management* 24(11), 4–15.

Sustainability Victoria. 2005. *Sustainability in Action: Towards Zero Waste.* Melbourne: Sustainability Victoria.

Watson, M., Bulkeley, H. and Hudson, R. 2008. Unpicking environmental policy integration with tales from waste management. *Environment and Planning C: Government and Policy*, 26(3), 481–498.

Williams, C.C. 2002. Why do people use alternative retail channels? Some case-study evidence from two English cities. *Urban Studies*, 39(10), 1897–1910.

Williams, C.C. and Paddock, C. 2003. The meaning of alternative consumption practices. *Cities*, 20(5), 311–319.

Williams, C.C. and Windebank, J. 2005. Why do households use alternative consumption practices? Some lessons from Leicester. *Community, Work and Family*, 8(3), 301–320.

Chapter 9

Build It Like You Mean It: Replicating Ethical Innovation in Physical and Institutional Design

Louise Crabtree

This chapter builds on research into challenges for innovation in developing, designing, building and occupying households with the dual aims of sustainability and affordability (see Crabtree 2006a, 2006b). How to replicate or foster innovation through the housing system was a key challenge identified in that work (see Crabtree and Hes 2009).

This research therefore focuses on issues related to disseminating innovation in housing tenure and governance forms, to sit alongside work on issues related to disseminating innovation in housing design and occupancy – specifically 'cohousing'. Cohousing has manifest as small-scale community-driven developments of around 30 or 40 households. Recently cohousing groups in the United States have partnered with affordable housing providers to more easily deliver affordability. This generates several issues of interest, including tensions between community-based and institutional imperatives, and how to scale up or transpose an ideologically conceived design agenda without prescription. My interest concerns design of institutional arrangements that can underpin innovation in affordability and sustainability, such as Community Land Trusts, and how to broaden these collaborative, hybrid structures and normalise these within housing systems. Physical and institutional design has a core role to play in facilitating (unconsciously) benign behaviour.

This work is part of a research trajectory exploring the role of physical and institutional design in enabling or facilitating sharing in cities and draws on radical democracy, feminist interpretations of home, complex adaptive systems theory and eco-city design. Each of these bodies of theory has been enunciated earlier (see Crabtree 2006a, 2006b) so will not be dealt with in any great depth here. For the purpose of this chapter it needs to be said that these theoretical frameworks share a concern with and focus on the embeddedness of the phenomenon under enquiry (the citizen, the home, the ecosystem, the city) within other systems and as an ongoing, contextual articulation of interacting forces and phenomena. With regard to sustainable or resilient cities or housing, consideration needs to be given not only to physical design concerns but also to the institutions by which physical spaces or systems are governed. Hence this work draws not only on design work that is striving to articulate physically resilient and flexible design, but also research into the management of complex adaptive systems, usually referred to as 'adaptive co-management'. Within housing research, work informed by complex adaptive systems theory has primarily focused on networks within systems of housing provision (for example Rhodes 2007) or on collaborative decision making in planning (for example Rhodes 2008). It has not yet focused on adaptive co-management on an ongoing basis (that is, after occupancy) or on the idea of the commons, which is experiencing a renaissance within resilience theory and complex adaptive systems theory (see Dietz, Ostrom and Stern 2003, Ostrom 1998).

This chapter therefore explores housing from a complex adaptive systems perspective and focuses on physical and institutional design as informed by theories of resilience and adaptive co-management. These theories offer intriguing insights.

Cohousing and Community Land Trusts Are...

Cohousing emerged as a design philosophy and practice in northern Europe in the 1970s. While cohousing has an expanding practitioner literature (for example McCamant and Durrett 1994, Meltzer 2005, ScottHansen and ScottHansen 2005) little attention has been paid to it by academics, although a themed issue of the *Journal of Architectural and Planning Research* (2000) did provide a start. Cohousing focuses on reducing individual household or unit size, complementing this with access to shared spaces in a common area or house. In its early days, such design was based on a desire for sharing resources and hosting events at a neighbourhood level and the social interaction that kind of activity fosters. As such cohousing was conceived primarily in terms of a concern with social issues, aiming as it did to develop community-based, community-oriented housing, interpreting and asserting community as a locally based forum in which to share resources, spend time and direct group efforts. During the 1980s and 1990s cohousing expanded into the United States and, to a very limited extent, Australia.

More recently – roughly at the same time as its expansion beyond Europe – cohousing has focused increasingly on the physical benefits of its design ethos. Some of these benefits were unconscious or unintentional, emerging spontaneously as a result of design outcomes or group dynamics once sites were occupied. These included reduced energy and water usage due to reduced unit size; reduced embodied energy through a reduction in duplication of some household appliances and tools; reduced car usage through car-free developments and car pooling among residents; reduced green waste through shared composting schemes; and the emergence of on-site food gardens. Cohousing developments now focus much more consciously on designing on the basis of reducing individual household consumption and establishing spaces for shared activities such as composting; the sharing of tools, appliances and cars; bulk buying; and food growing. Further, many cohousing developments now incorporate low energy design features such as solar passive design; effective levels of insulation in ceilings and walls; use of thermal mass; sustainably sourced materials; double glazing or thermal glass; on-site water collection, treatment and reuse; and waterless toilets. The uptake of these has not been uniform or ubiquitous, varying according to groups' aims and resources and local circumstances, including regulatory parameters and the availability of materials, information and skilled labour.

Irrespective of the extent of energy and water concerns in design, cohousing retains at its core the development of compact individual units or houses focused on and complemented by group facilities provided in a common house or space. At their most basic these facilities include a kitchen and meals/meeting room and a bathroom, as most cohousing groups organise shared meals on a regular basis, usually weekly. Attendance at such meals is rarely compulsory; however most cohousing groups establish rosters for cooking, setting up and cleaning in which individuals are required to participate at a given level. Common additional shared facilities include a laundry and guest room(s). Other facilities and uses vary from development to development, but have included workshops, offices, cool rooms, children's play rooms, photographic dark rooms, libraries, film rooms, sheds, food storage areas and collective food distribution points. Usage and maintenance of these spaces is usually managed by user groups, which may become semi-formal committees. Numerous cohousing groups find that usage of these spaces also extends beyond the resident group, with surrounding neighbours attending shared meals and events, or using shared spaces, most often for meetings of local community groups. Further, design aims to foster spontaneous interaction and exchange through shared points of access and connection between individual and community spaces.

The basis of ownership in cohousing varies between communities and between countries. In the United States a lot of cohousing is condominium based, with legal and ownership arrangements roughly parallel to strata title in Australia. In Australian strata title householders hold direct title to their unit and access to the shared facilities. Much cohousing is owned as housing cooperatives, where residents own shares in a cooperative corporation which owns the housing and the land. Shared ownership grants the shareholder the right to reside in a unit of housing and access to the shared facilities. In the United States most cooperatives operate as market cooperatives in which shares sell at a market rate. However some operate as limited equity cooperatives, in which the value of shares is indexed in some way to retain the affordability of shares over time.

Increasingly the social agenda of cohousing has broadened to include issues of the upfront and ongoing affordability of housing. Due to the high level of design consciousness much cohousing becomes unaffordable either by the time of completion and hence at first purchase, or over time as the desirability of the development increases due to the on-site culture and activities, or due to emergent benefits such as improved thermal comfort due to environmentally sound design. In addition to issues of affordability, cohousing has protracted development times and high failure rates; typically cohousing developments take in the order of ten years to complete, with a completion rate of one in ten start-ups. This can in large part be attributed to community-driven design and development processes, which traditionally occur through the formation of a core group of interested individuals who then become the driving force in maintaining a collaborative design process. Not surprisingly, many groups do not – although some may and do – possess property development, architectural, project management or group facilitation skills. Furthermore most meetings and development usually occur in addition to, not as, the individuals' primary source of income, placing fairly hard and fast constraints on how much time and effort individuals can commit and on the number of individuals who can and will participate. Many cohousing groups, designers and advocates are consequently investigating mechanisms to provide affordable housing and to bring cohousing developments to completion faster, with less demand on future residents.

Paralleling the expansion of cohousing over the past 30 years has been the articulation and expansion of Community Land Trusts (CLTs) in the United States. Initially borne out of the civil rights struggles of the southern states, CLTs aim to provide and maintain housing that is affordable in the long term, with the ideal of holding property in perpetuity. Traditionally CLTs have been private, non-profit organisations holding title to land for the purposes of affordable housing and community benefit. Property holders then hold title to any improvements on that land, which may be anything from the ground up: crops, sheds, houses, units, businesses and so on. Improvements can be held on a rental or ownership basis, and can be held by individuals, cooperatives, other non-profit groups or companies. The relationship between a CLT and the property holder is spelt out in the ground lease, which grants full usage rights to the land and sets out restrictions regarding occupancy, usage, eligibility, resale values or rents, maintenance and other requirements as deemed fit by the CLT. Many CLTs have a tripartite board structure of equal parts resident members, non-resident members and public interest representatives such as government, local businesses, lenders, chambers of commerce, architects, planners, community groups and charities. This structure is intended to balance interests on the board so that tenants cannot vote away affordability conditions and gain windfall profits; the NIMBYism[1] often faced by affordable housing developments cannot easily take hold; and state directives cannot dominate. In addition most CLTs build in multiple mechanisms to prevent the CLT from veering from its core aims or to provide fall back mechanisms to safeguard any public subsidisation.

CLTs hold title under mortgagee homes, rental homes, housing cooperatives, shelters and boarding houses. Parallel programs and activities have emerged on CLT properties, such as community food gardens, child care, aged care,

---

1   'Not In My Back Yard', or NIMBY, refers to an attitude of opposition by existing residents in an area to proposed new developments, frequently encountered when affordable or social housing projects are being established.

businesses, charities, medical centres and farms. Perhaps not surprisingly cohousing developments are increasingly looking to CLTs to not only provide affordability but further reflect the community ethos of cohousing, because CLTs at their core manifest and maintain ongoing relationships between householders, their surrounding community and the broader public. From a concern with the material geographies of household consumption, this relationship between cohousing and CLTs is an irresistible lure. The issue of broadening the uptake and scale of both cohousing and CLTs is thus of great interest, and in the United States both CLTs and cohousing groups are currently interested in, and grappling with, further expansion.

Upscaling Is...

Complex adaptive systems theory and adaptive comanagement are based in understandings of the embeddedness and contextuality of systems. Complex adaptive systems are described by Per Olsson, Carl Folke and Fikret Berkes (2004: 76) as systems:

> in which properties and patterns at higher levels emerge from localized interactions and selection processes acting at lower scales and may feed back to influence the subsequent development of those interactions. They are characterized by nonlinear relations, threshold effects, historical dependency, multiple possible outcomes and, limited predictability.

Cohousing reflects an understanding of the individual home as an embedded and multi-scalar entity (Crabtree 2006b): the home is embedded in, responding to, shaped by and shaping, systems of resource extraction, production, circulation and disposal, as well as the circulation of norms and practices regarding living patterns, design, behaviour, legislation, finance and tenure. It is very hard to identify where a cohousing development 'starts' and 'ends', or its scale, given the conscious connections made between individual homes, the cohousing development and the systems in which the development is embroiled. In this context the idea of replicating cohousing is not a matter of upscaling in any linear sense, but something far messier and rhizomatic.

Cohousing advocates realised early on that access to sympathetic designers was crucial to the development of housing deemed appropriate for cohousing. The San Francisco-based architectural duo Katherine McCamant and Charles Durrett have been fundamental to the expansion and dissemination of cohousing, both through their practice and through the publication of the core cohousing tract, *Cohousing: A Contemporary Approach to Housing Ourselves*, in 1988, with expanded versions in 1994 and 2004. *Cohousing* not only outlined the design philosophy behind the practice, but provided design plans, photos and in-depth histories of existing cohousing developments. Brad Gunkel, an architect with

McCamant and Durrett, spoke of the challenges of getting cohousing developments completed (Brad Gunkel 28 April 2008, personal communication). These included the developers' view that community-controlled development is a disincentive, meaning there were only a handful of repeat developers. Consequently it was felt that a more efficient model for replication and development was needed than the traditional cohousing development process. Katherine McCamant and Charles Durrett's practice has thus recently expanded to include partnership with a local affordable housing provider, Affordable Housing Associates (AHA), to undertake an affordable rental cohousing development, Petaluma Avenue Homes.

At the time of research (2008), the Petaluma Avenue project brief was for the construction of 45 homes for purchase and rental by low-income residents built around common facilities on one hectare of land in Sebastopol, California (see Figure 9.1). The 290m^2 common house would contain a dining room, kitchen, lounge, kids' room, laundry and computer room. The project was initiated by the city, which dedicated the land for the project. All residents were to be subject to an income and assets test and as per the state's fair housing law, must be selected via a lottery. The homes were to be a mix of one and two bedroom units and three bedroom townhouses. The development process was planned as a hybrid between cohousing's traditional, community-controlled development and planning process and the legally required eligibility and lottery process.

Typically cohousing projects develop through the core team meeting to develop and refine a site plan according to the intended resident group's requirements, aims and objectives. However, as the intended residents of Petaluma Avenue could not be guaranteed because of the lottery system, at the time of the project's inception there was no guaranteed resident base to drive the design process. To address this AHA hired a cohousing consultant for two years to act as a facilitator to run the 'cohousing club', which involved regular social and information sessions aimed at creating a group culture despite the lack of specific design input. The consultant was also to keep working on-site after completion of the buildings. The consultant would work with the cohousing club to perform tasks usually undertaken by a cohousing core group, such as drawing up meals plans and garden proposals. Further, an advisory committee was to be formed from the potential resident group and existing neighbours, with the aim of providing some community direction to the design process and giving both potential residents and the surrounding community a sense of ownership of the process (see Figure 9.2 for an overview of the unique development process). However it is not hard to imagine that buy-in to the group by prospective residents would be limited, as it would be an altruistic individual indeed who would spend hours of voluntary time on a project and community they had no guarantee of living in. It was also unclear what the roles and activities of surrounding neighbours would be once the steering committee wound up. In 2008, as per Figure 9.2, the cohousing facilitator provided the only guaranteed continuity across design and occupancy of the site.

Figure 9.1 Promotional Image of Petaluma Avenue Homes, Sebastopol, California. Prior to Completion
One and two bedroom apartments on the left face three bedroom town houses on the right with the common house in between. Design by McCamant & Durrett Architects 2008

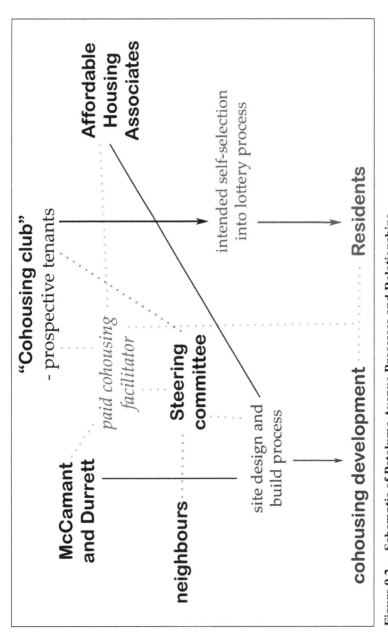

Figure 9.2   Schematic of Petaluma Avenue Processes and Relationships
Solid lines represent processes and dashed lines relationships. Design by McCamant & Durrett Architects 2008.

To address this it was hoped that the cohousing club would serve a second purpose of effectively filtering a core group of pre-qualified future residents via self-selection into the lottery process; again, however, none of that was guaranteed.

With housing affordability in the area quite compromised, it was likely that any new affordable housing being made available would attract a substantial pool of eligible applicants, most of whom might have little knowledge of or interest in cohousing. This raises issues regarding coordination and management of facilities and activities on-site after completion, which AHA hoped to address through the continued activities of the consultant for some time after completion. However it also raised the issue of whether individuals need to be consciously interested in cohousing as such; ideally, individuals will not have to be, simply being attracted to on-site facilities and activities. These do however require a certain core level of maintenance and facilitation on an ongoing basis, a point to which we will return later.

The Petaluma Avenue site was designed with kitchens and balconies facing central, car-free courtyards, in a design that aimed to maximise the potential for informal daily interaction (Brad Gunkel 28 April 2008, personal communication). This is a central goal of cohousing; architects Katherine McCamant and Charles Durrett found that less proactive cohousing design (for example allowing multiple individual points of entry to the site) frequently undermined the number of spontaneous community and social exchanges on-site and weakened the sense of community. While this may sound unnecessarily prescriptive, cohousing design has spent a lot of time and energy addressing the balance between individual household and shared space, aiming to house individuals in a community, but not in each other's pockets or faces.

With this in mind we can call up the spirit and reality of the Radburn estates, similarly built along design principles derived from the Garden Cities movement and ambitiously deployed in social housing contexts throughout the 1970s in both the United States and Australia – but, perhaps crucially, without resident input. Radburn estates shared with cohousing a focus on central, car-free spaces and pedestrian access. However they did not as a rule provide community facilities, or access to processes and spaces for engagement. Such design has been variously described from being of high to being of no significance to consequent social problems that arose on some of the Radburn social housing estates. One of Sydney's most notorious and troubled Radburn estates, East Fairfield, was subject to contested demolition in the late 1990s. The report into that estate which proposed its demolition assessed the situation as 'an excellent example of the schism ... between the architect's vision of a pedestrian-focused, village style community and the reality of how low-income households organise and behave in such a space' (NSW Audit Office 1998a: 12).

However it also stated that 'there is no simple connection between environment and behaviour' (NSW Audit Office 1998b: 32). The report referred to the *NSW Bureau of Crime Statistics and Research Report on Public Housing and Crime in Sydney*, which had similarly suggested 'there is no relationship between crime and estate design' (NSW Audit Office 1998c: 38). Kathy Arthurson's article on the discourse involved in justifying the demolition of East Fairfield states that:

> the perception of East Fairfield 'as disastrous and ugly ... is worse than the problem itself.' This image was propagated by people who did not live in the estate ... At Claymore estate ... there were similar problems to East Fairfield with the Radburn design and social amenity of the estate. In 1995, when the community development programme commenced, the estate had the dubious reputation of being the worst suburb in NSW. Since then, the situation has improved considerably through undertaking minor physical design changes to block off walkways, implementing night patrols to address crime, and residents developing, amongst other initiatives, a community-run launderette and coffee shop. (Arthurson 2004: 262–3, 267)

While this chapter is not concerned with examining the disparate fates of East Fairfield and Claymore in any depth, the role of design in these histories is fascinating and relevant. Earlier recommendations that did not include demolition suggested that East Fairfield could be assisted through community initiatives and minor design amendments similar to those later deployed successfully at Claymore. In both instances the primary design issue of note was the ability of individuals to elude police through substantial networks of internal pedestrian paths. As stated above, architects McCamant and Durrett also highlighted multiple entry and exit points as undermining community, and consequently now design developments with only one or two entry points. A minor issue at East Fairfield concerned the fact that the estates looked different to surrounding housing as the focus of the houses was on internal courtyards and, ironically, because they were slightly offset from the road for maximum solar gain. This difference was perceived as adding to the stigmatisation of the estate.

This raises several issues for consideration. Firstly, there is a clear need for both social infrastructure and appropriate physical design. That is, the role envisaged for the addition of community facilities in addressing social issues at Claymore suggests that cohousing is on the right track in building in such spaces at the onset: it is not enough to focus houses on a central point if there is nothing at the centre. Moreover ongoing community organisation and participation were highlighted at both East Fairfield and Claymore as key components in redressing social issues. That is, once community spaces are built, they must be used and maintained on an ongoing basis, which may be an issue for Petaluma Avenue. Traditionally cohousing communities establish participation at the outset and then rely on the terms of residency to ensure participation. This can provide a range of abilities for fostering participation, ranging from self-selection of individuals prepared to participate, through agreed rosters, to legal mechanisms for eviction (even if evictions rarely occur readily or easily). At its inception, Petaluma Avenue only had self-selection, and fair housing law means further eligibility criteria could not be added.

At this stage it is worth exploring the issues faced by CLTs in increasing their size, activities and scope as a result of their great emphasis on upfront and ongoing involvement of residents, neighbours and the broader public. In addition to the tripartite board structure outlined above, regular building, fundraising and social events maintain core relationships between the parties represented on the board. Potential and actual CLT residents are the subject of most of the CLTs activities, as CLTs educate and prepare prospective residents and work with residents to redress any financial issues that emerge. Most CLTs also encourage residents to establish sub-committees to manage various programs and to act as training and mentoring mechanisms for members skilling up for directorship roles. Additionally CLTs are often used as de facto lobby groups, with residents and neighbours approaching CLTs with issues they want taken to local government. However while CLTs pride themselves on this civic role, Robert Silverman (2009) refers to this in other community-based housing organisations (CBHOs) as the result of these organisations being mistaken for state bodies, despite their community bases and ambitions, a possibility not yet raised within the CLT literature. That literature has tended to focus on documenting the history and variations of the model or on its role in providing affordable housing (for example Davis 2010, Paterson and Dunn 2009). An issue that is raised is whether the slow drift in the focus of CLTs away from land rents (based on Henry George's single tax model) towards the promotion of homeownership via resale, has been a concession to dominant interpretations of property and homeownership (John E. Davis 23 March 2009, personal communication). Further, it has been argued that focusing on CLTs purely from a perspective of housing affordability – especially when focused on conceptually and politically loaded forms such as mortgagee ownership – detracts from the model's potential for decentralised, participatory and ongoing control over planning and development, a point to which we will now turn.

Historically, and similar to cohousing, CLTs have mainly been driven by community-based activists working in concert with supportive individuals and organisations, whether public or private or community based. Increasingly however, CLTs are being established by either municipal government or established affordable rental housing providers, analogous to Australia's community housing providers. This has primarily been in order to create CLTs faster and in greater numbers: currently the largest CLT has stewardship of around 2,600 units of housing scattered across three counties in Burlington, Vermont. This stock has taken 25 years of solid effort to build up, including through a merger with an affordable rental housing provider. In contrast the city of Irvine, California, formed a CLT in 2007 to develop 10,000 homes over ten years with a grant of US$250m; Chicago, Illinois is planning a similar scale of implementation.

This has clear advantages for the rapid delivery of a greater number of new, permanently affordable homes. However issues and debate are emerging about governance and the philosophy behind the formation of these CLTs. Given the hefty investment by government agencies, the boards of these CLTs will still be tripartite, but with the individuals of all three parts appointed – or at least vetted

– by the relevant government body. This is in contrast to the traditional process of election of CLT boards by CLT members only; it remains to be seen whether this is significant. It does seem conspicuously heavy-handed, as some CLTs already build in mechanisms whereby the state has the right to intervene should the CLT stray from its stated principles or whereby title reverts to the state should the CLT fail financially. This heavy-handedness has led to speculation as to whether this is an unfortunate consequence of ignorance about the protective mechanisms available, the reflection of an assumption that communities are unable to effectively govern organisations with large property portfolios, or a strategic manoeuvre to keep the terms of development firmly defined by the state. It would seem straightforward to build the required safeguards into these CLTs without interfering in the processes of board appointment; this could include training and mentoring board members to a required level of competency as part of the nomination process, as already implemented by Dudley Neighbors, Inc. in Boston, Massachusetts.

Design Is...

For now we will assume this state heavy-handedness to be benign and focus on the opportunities to enact governance via these larger institutions. Robert Silverman (2009) documented the impact of increasing size on community perception and development of CBHOs in the United States. Silverman (2009: 13) states that:

> The adoption of relatively expansive boundaries ... diluted the access of low-income, minority residents to decision-making in the organisations. As a result, while institutions and the middle class were well represented on governing boards, none of the executive directors indicated that renters, the poor or other indigent groups were highly visible.

He further states:

> In essence, disincentives existed for CBHOs to pursue community organising and advocacy work, since rewards came from conforming to decision-making processes that were centralised ... For the scope of citizen participation to expand, [local administrators'] roles would have to shift to a focus on facilitating and monitoring systems designed to expand grassroots control of local community development. (Silverman 2009: 19, 22)

This highlights numerous issues of relevance in increasing CLTs' portfolios and which echo the role of the cohousing facilitator in Petaluma Avenue. The primary issues relate to representation and the maintenance of community participation. Cohousing faces the additional challenge-cum-opportunity of integrating these issues through physical and institutional design.

Relying on design alone to augment sustainable behaviour is well-intentioned but potentially flawed inasmuch as there are dangers of prescribing behaviour, that is, where design might be seen as didactic or dictatorial. Relying on design alone does not account for or accommodate the vagaries of human behaviour, as seen in many of the chapters in this volume (see Dowling and Power, Hawkins and Race, Paling and Winter). Consequently, in addition to advocating for unconsciously environmental design (design that is environmentally benign without people knowing it or having to think about it, such as low flow taps), recent design activism is focusing on the role of designers as ongoing facilitators and amplifiers of community-based initiatives, ideas and agendas, such as the Sustainable Everyday Project (Sustainable Everyday Project 2010). This represents a move away from the designer as arbiter of taste, or architecture as egoistic statement, towards design as an ongoing service driven by and answerable to community objectives.

This echoes the emerging description of CLTs as 'the developers who don't go away', that is, CLTs design and build, but are also involved as partners, supports and lobbyists on an ongoing basis, governed by a board aiming to be accessible, balanced and representative. The ongoing presence of CLTs and their perpetual interest in properties is generating much innovation, both in terms of design and build qualities, and institutional arrangements. Firstly, many CLTs are building high quality homes with a high level of energy and water saving features and an increasing degree of universal design (ideally, design for all abilities and ages). Secondly, CLT ground leases are being used to establish who is responsible for what type and level of maintenance, as well as frequency and type of property inspections. These clauses are the result of years of trial and error and carry the dual aims of respecting residents' (frequently hard-earned) right to quiet enjoyment of their home and maintaining the affordability and condition of those homes. This is the result of having multi-scalar boards that combine the interests of various parties, including residents, neighbours and government or business agencies.

This role of the CLT as steward and facilitator echoes Silverman's assessment of the role of local administrators, the role of the cohousing consultant at Petaluma Avenue, and the role of designers as advocated and practised by groups such as Grupo DESIS (see Grupo DESIS 2009, Sustainable Everyday Project 2010). This ultimately establishes design for sustainability as a hybrid of physical system design, aiming for as little conscious effort or perceived 'weirdness' as possible, and ongoing facilitation and amplification of residents' objectives. As per Silverman's statement, there is a new role for local administrators in this such that communities are not left to sink or swim according to their ability to establish and maintain sustainable systems. Rather, service delivery needs to look at ways of establishing and supporting mediating spaces between centralised provision, support and monitoring, and decentralised design and control. Similar to cohousing design, these are systems which blur the 'scale' of the household, bringing multiple interests to bear on a system of provision on an ongoing basis.

Moreover tenure and occupancy forms may also need to shift into new terrain to foster the time and space for participation. An interesting example is the 'city cousin' in Food Connect Brisbane. Food Connect Brisbane is an evolution of traditional community supported agriculture schemes in which urban subscribers pay a fee to receive a weekly delivery of organic, seasonal produce, delivered to a number of distribution points on a given day of the week. The subscription fees cover basic overheads, with the bulk of them going directly to primary producers, who earn higher returns than when dealing with mainstream forms of distribution, while keeping prices below those in supermarkets. The distribution points are voluntarily staffed by the 'city cousins' – individuals who make time and space available to receive boxes and oversee their collection in return for discounted subscription fees. The Petaluma Avenue facilitator might be seen as playing an analogous role in cohousing, that is, training and mentoring a core group of residents who then become the ongoing core team responsible for coordinating on-site events, systems, groups and activities in return for discounted rent, ground lease or mortgage payments, subsidised by and reporting to, a CLT or other housing body. As per Silverman, local administrators here would be responsible for supporting a network of such individuals and monitoring outcomes and issues within their jurisdiction.

Such decentralisation is starting to occur within social housing in Australia but is in its early stages. Community housing providers (CHPs) are currently receiving title transfers to stock such that the sector will become the primary growth engine in social housing provision. However the sector is struggling with issues of low resident participation, primarily due to a lack of genuine opportunities for this. Whereas the embryonic cooperative housing sector relies on and demands involvement, and hence experiences relatively high levels of participation, many CHPs struggle in this regard. Furthermore many CHP boards are wary of tenant involvement in governance and are unaware of mechanisms for enabling tenants to become effective directors. Establishing systems for genuine input and control that are structurally supported, readily accessible, transparent, accountable to multiple parties, and easy to take part in, would seem a timely proposition.

Governance Is...

Work on adaptive co-management has emerged from a concern with how best to manage complex adaptive systems, and such work highlights that the governance structures best able to effectively manage complex adaptive systems need to be flexible and iterative and to combine knowledge from various bodies. Adaptive co-management systems need to be able to receive, decipher and respond to feedback, and draw on knowledge from multiple scales to learn over time. Given that adaptive co-management is multi-scaled and always contingent, how can the practical outcome of more cohousing be pursued and what governance mechanisms might underpin this?

Ultimately the facilitation of innovative housing design should take some of the pressure off core individuals and make innovation easier. As such, skilled designers should be in place from the outset, facilitating and amplifying design and group development during construction and after occupancy. A key issue is how to identify and develop the core resident group during this process, particularly when its members are bound by issues of eligibility and equitable access. It may be that individuals willing to act as core organisers will become a separate category of eligible applicants, independent of income levels.[2] This may address the eligibility issue, but it does not address the issue of how prospective residents in full employment may participate prior to residency. It may be that the city cousin analogue has to be established as part of the design process, with a degree of remuneration for time, which then evolves into the full relationship upon residency.

This could represent a substantial reinterpretation of existing relationships between housing providers and those on wait lists, upsetting the welfare recipient applecart by paying community members to co-design and co-manage affordable housing. It reflects adaptive co-management in combining resources and knowledge from various agencies, providing structural support (whether funds, training or other supports) to decentralised and, hopefully, responsive governance. The key is to establish and support flexible governance mechanisms operating with relatively hard and fast objectives regarding social and environmental justice; in this the role of the provider shifts to one of ensuring and supporting arenas for ongoing negotiation and management of the terms of development and for providing structural support for locally based, accessible and transparent systems. This is vital if the process is not to become a co-option of potentially marginalised households into unsympathetic state agendas. Lastly this is not an assertion that all housing developments should self-manage; however, all should have the ability and support to do so, plus access to broader support if there is not the capacity or desire for self-management. This is the basic framework of the Swedish cooperative housing sector, which represents 15 per cent of that country's housing stock. In this, a nested hierarchy of cooperatives means that individual cooperatives that are unable or unwilling to self-manage have access to overarching cooperatives that can take on these roles. Further, each overarching cooperative is owned by the member cooperatives which it oversees, thus increasing accessibility and accountability.

There is the final issue of the normalisation of innovative infrastructure, such that sustainable systems are in place and maintained by individuals willing, able and supported to do so, but not having to expend too much effort to do so. An analogy is found in Food Connect Brisbane's system, which is maintained by

---

2   Within the requirements of housing organisations: for example some Australian CHPs can house people in different income bands ranging up to roughly A$75,000 for a household of five; housing cooperatives in New South Wales can have a mix of 65 per cent of tenants eligible for public housing and 35 per cent on incomes above this, with no maximum income limit; CLTs in the United States vary their eligible income range but this can go up to 120 per cent of median income of each given area. Community Land Trust residents do not become ineligible should their incomes rise once they are housed.

and rewards interested people willing to be the core, while servicing some 1,000 subscribers via a website and direct payment options. The issue is to similarly structure and replicate innovative housing design and tenure forms such that the various physical and institutional systems can be maintained as the invisible infrastructure underpinning them. As with physical cohousing design, which houses the individual in community, use of and access to these institutional systems of ownership and governance should seem effortless and inconspicuous. This ultimately requires resources that, in the absence of systems like the Swedish cooperative housing sector, will most likely be key contributions from the state. It may also be that as CHPs expand their asset bases, skills and tenure forms, they will be able to take on this facilitating and resourcing role, ideally underpinned by a core of guaranteed funding from the public purse, but also increasingly able to internally cross-subsidise supported housing through broadening their tenant, usage and tenure bases.

Such a reinterpretation also suggests revised and expanded roles for utility providers in the establishment and maintenance of decentralised systems. Taking the example of on-site water collection, treatment and reuse, these are currently generally managed by resident populations, suggesting perhaps unrealistic demands on householders' time and abilities. Similar to ongoing on-site resident facilitators, utility provision could expand to provide centralised service provision through the training and support of residents who, according to abilities, availability and interests, could undertake on-site utility maintenance. As with the types of hybrid systems represented by CLTs and city cousins, perhaps major maintenance falls to the utility company, with interested residents being trained in minor ongoing maintenance in return for reduced service fees. This again unsettles traditional relationships between 'providers' and 'consumers', and would require proactive, or at least adaptive, utility companies (see also van Vliet, Chappells and Shove 2005 on decentralised utility provision). The current inability of energy providers to consistently accommodate or reimburse decentralised electricity generation in Australia highlights a current lack of institutional capacity for decentralisation. Such challenges, though, will only become more pressing as more decentralised systems come into play.

In All…

This chapter's core concern has been with the replication of the conscious design focus of environmentally designed cohousing, without requiring that a majority of residents be recruited to its development and management. That is, the hope is that larger developments may be brought into being which set up hybridised physical, ownership and managerial infrastructure that blur producer–consumer and provider–recipient binaries, and which enable socially and physically just systems to function on the basis of resident participation to the extent that residents are willing, able and supported to do this. Currently the development and maintenance of sustainably designed cohousing require efforts that verge on the superhuman; they should (and can) be run-of-the-mill.

Such systems require and combine resources and skills from multiple scales on an ongoing basis, echoing lessons and models of the adaptive co-management of complex adaptive systems (see Olsson, Folke and Berkes 2004). In such systems, many of the traditional binaries of citizenship and governance, such as private–public and empowerment–co-option, become obsolete, as the key objective and outcome is to establish hybridised, practical, accessible and representative fora for the ongoing definition, contestation, co-management and monitoring of systems and processes aiming to address and combine social and physical justice. It is perhaps not surprising that such a description echoes the languages of radical democracy and feminist reinterpretations of home. It is key that such systems and processes are open and accountable to their publics and that participation not be interpreted or implemented as annexation or outsourcing; rather, that hybridity in governance and design enable and enhance contestation and negotiation over the terms and forms of development and living.

Acknowledgements

My thanks to Brad Gunkel of McCamant and Durrett for his time, materials and thoughts and to Nicole Cook for always listening and providing feedback.

References

Arthurson, K. 2004. From stigma to demolition: Australian debates about housing and social exclusion. *Journal of Housing and the Built Environment*, 19(3), 255–270.

Crabtree, L. 2006a. Sustainability begins at home? An ecological exploration of sub/urban Australian community-focussed housing initiatives. *Geoforum*, 37(4), 519–535.

Crabtree, L. 2006b. Disintegrated houses: exploring ecofeminist housing and urban design options. *Antipode*, 38(4), 711–734.

Crabtree, L. and Hes, D. 2009. Sustainability uptake in housing in metropolitan Australia: an institutional problem, not a technological one. *Housing Studies*, 24(2), 203–224.

Davis, J. 2010. *Common Ground: the Community Land Trust Reader*. Cambridge, MA: the Lincoln Institute of Land Policy.

Dietz, T., Ostrom, E. and Stern, P. 2003. The struggle to govern the commons. *Science*, 302(5652), 1907–1912.

Grupo DESIS 2009. *Grupo DESIS* [Online]. Available at: www.ltds.ufrj.br/desis/english/index.htm [accessed: 19 August 2009].

Journal of Architectural and Planning Research (2000). 19(2).

McCamant, K. and Durrett, C. 1994. *Cohousing: A Contemporary Approach to Housing Ourselves*. Berkeley, CA. Ten Speed Press.

Meltzer, G. 2005. *Sustainable Community: Learning from the Cohousing Model*. Victoria, BC: Trafford Press.

NSW Audit Office, *see* New South Wales Audit Office

New South Wales Audit Office 1998a. *Redevelopment Proposal for East Fairfield (Villawood) Estate – Background* [Online]. Available at: www.audit.nsw.gov. au/publications/reports/performance/1998/villwd/backgrnd.htm [accessed: 14 August 2009].

New South Wales Audit Office 1998b. *Redevelopment Proposal for East Fairfield (Villawood) Estate – the Reasons for Demolition* [Online]. Available at: www. audit.nsw.gov.au/publications/reports/performance/1998/villwd/reasons.htm [accessed: 14 August 2009].

New South Wales Audit Office 1998c. *Redevelopment Proposal for East Fairfield (Villawood) Estate – Appendices* [Online]. Available at: www.audit.nsw.gov. au/publications/reports/performance/1998/villwd/appendices.htm [accessed: 14 August 2009].

Olsson, P., Folke, C. and Berkes, F. 2004. Adaptive comanagement for building resilience in social–ecological systems. *Environmental Management*, 34(1), 75–90.

Ostrom, E. 1998. Scales, polycentricity and incentives: designing complexity to govern complexity, in *Protection of Global Biodiversity: Converging Strategies*, edited by L. Guruswamy and J. McNeely. Durham: Duke University Press, 149–167.

Paterson, E. and Dunn, M. 2009. Perspectives on utilising Community Land Trusts as a vehicle for affordable housing provision. *Local Environment*, 14(8), 749–764.

Rhodes, M. 2007. Strategic choice in the Irish housing system: taming complexity. *Housing, Theory and Society*, 24(1), 14–31.

Rhodes, M. 2008. Complexity and emergence in public management. *Public Management Review*, 10(3), 361–379.

ScottHanson, C. and ScottHanson, K. 2005. *The Cohousing Handbook: Building a Place for Community*. 2nd Edition. Gabriola Island, BC: New Society Publishers.

Silverman, R. 2009. Sandwiched between patronage and bureaucracy: the plight of citizen participation in community-based housing organisations in the US. *Urban Studies*, 46(1), 3–25.

Sustainable Everyday Project. 2010. *Sustainable Everyday Project* [Online]. Available at: www.sustainable-everyday.net/SEPhome/home.html [accessed: 10 September 2010].

van Vliet, B., Chappells, H. and Shove, E. 2005. *Infrastructures of Consumption: Environmental Innovation in the Utility Industries*. London: Earthscan.

Chapter 10

Rethinking Responsibility? Household Sustainability in the Stakeholder Society

Andy Scerri

I think a lot of money can be spent on promotional campaigns, but if the message doesn't 'sink in' for some people, these campaigns won't work. Thanks!
Response to City of Melbourne Residential Sustainability Questionnaire, April 2009.

Introduction

Understanding citizenship as a status and a set of practices, this chapter examines citizens' beliefs and values in relation to local government policy for household sustainability. I first outline my analytical framework and then develop some explanatory propositions concerning citizenship in liberal democracies. I describe 'stakeholder' citizenship as emergent over recent decades and explain its relationship with policy for sustainable development. My key proposition is that official policy for achieving household sustainability is aimed at involving householders as stakeholder citizens. Such policy promotes individual self-responsibility and personal capacity building within a growth-oriented economy as the best way to achieve the sustainability goal.

In subsequent sections I use this scene-setting discussion to explore Kersty Hobson's argument that one side effect of stakeholder-oriented sustainable development policy is the opening up of discursive traps. These draw householders' attention to the ineffectual nature of individuated responses to climate change while highlighting 'what [is] wrong with society' (Hobson 2002: 112). I then use Hobson's work as a point of departure for an analysis of questionnaire responses designed to understand householders' attitudes and values in relation to local government action on household sustainability in Melbourne, Australia. Based on this analysis, I argue that householders are rejecting official claims that 'rational' sustainable consumption choices and self-regulatory approaches will achieve the kinds of changes that sustainable development necessitates. I contend that people associate household sustainability with a more general conception of sustainability as a virtue: householders see 'greenness', in opposition to 'growth', as the most important way of defining the common social good. The key implication of my argument is that, although largely problematic as material contributions to household sustainability – the 'indicators' continue to go in the wrong direction – sustainable consumption and eco-efficient strategies need to be reconceptualised as tools for promoting value change and redefining sustainability as a social problem.

Note on the Approach

My conceptual approach draws on the critical pragmatism developed by Luc Boltanski, Laurent Thévenot and others. Boltanski and Thévenot build upon work by Bruno Latour and Michel Callon, as well as Louis Dumont and Paul Ricoeur (Boltanski and Thévenot 2006: 20). At a less abstract level, I view citizenship as a narrative of change over time. I draw on Joaquin Valdivielso's recent discussion of T.H. Marshall's seminal 1950 work on citizenship as the practical fulfilment of a liberal democratic ideal. Citizenship describes individuals' rights and duties within a given polity (Marshall 1998, Valdivielso 2005). It is situated within society at the nexus of politics – organising the rules for life held in common – and culture – the production and dissemination of meanings that cohere and resonate over time.

In view of this I approach citizenship in materialist terms as a set of practices and discourses *and* as something two-dimensionally 'ideological' (Dumont 1986, Chiapello 2003). It represents both a political-ideological 'distortion and dissimulation' of power and a cultural-ideological whole set of representations that 'facilitate social integration and identity preservation'. That is, citizenship is not merely a legal status but a way of life. It provides a framework for public debate over the legitimacy of the values, norms and rules that apply in society, while also informing what it means to be part of that society. Citizenship is the contingent historical product of a (relatively) peaceful 'co-struggle' between and across political social forces (Karagiannis and Wagner 2008: 324).

In this view the practices and discourses that social actors create represent reasons or 'justifications' for such actions. These justifications draw on an established cultural ideology. However in order to be acceptable as valid reasons for action they must appeal to the general interest or common good. That is, they are politically charged. Given that politics is fraught with disagreement about how best to achieve the common good, there exist within society different and often opposing 'justifications' for action (Boltanski and Thévenot 2006). Indeed different actors hold different and often incompatible ideas and beliefs about how best to achieve the common good. This tension between reasons for acting in different ways to achieve the common good puts actors at risk of being publicly exposed as politically distorting and dissimulating social representations in ideological ways.

In these terms particular practices and discourses can appear unworthy as contributions to the common good because they are grounded in a different order of justification. It is here that 'compromise situations' emerge (Boltanksi and Thévenot 2006: 277), as practices and discourses justifiable by reference to one order of justification overlap and are made compatible with those of another. A 'compromise' arises when two different conceptions of the common good, and how best to achieve them, maintain support within a single polity. For example during the Cold War advocates of socialism and conservatism established just such a 'welfare state' compromise between capitalistic-nationalist and collectivist-paternal values. Such compromises remain unstable however, as neither of the opposing orders of worth is altered. That is, there exist significant practical and discursive tensions between the different courses of action that an order of justification supports. In compromise situations, an 'intentional proclivity towards the common good' is maintained without addressing political tensions directly: 'without [attempting] to clarify the principle upon which ... agreement is grounded' (Boltanski and Thévenot 1999: 374). In situations characterised by such compromise, attempts to return to one or another order of justification, or to map out a new order, are marginalised.

In the following sections I apply this analytical schema to examine how just such a compromise between two orders of justification is being formed and re-formed around the sustainability issue. In this view policy for sustainable development represents a compromise between an order of worth based in *growth*, referring to the common good in terms of price (exchange-value) within growth-oriented markets, and a *green* order of worth, referring to the common good in terms of a sensitivity to the limitations that the environment places upon society (Thévenot, Moody and Lafaye 2001). Of course, social movements and critical academics have long argued that the sustainable development compromise is flawed and inconsistent (Daly 1999, Hutchinson, Mellor and Olsen 2002, O'Connor 1996, Perelman 2003). However what my analysis of responses to the questionnaire demonstrates is that householders themselves express dissatisfaction with the personal responsibility and institutional voluntarism central to it. I contend that it is in the context of stakeholder citizenship that the compromise between economic growth and green values is being weakened.

Stakeholder Citizenship and Sustainability as a Virtue

Elsewhere I have argued that stakeholder citizenship emerged as the prevailing status and practice of citizenship in the Anglo-American liberal democracies as neo-liberal reformism gradually dismantled the welfare state in the last decades of the twentieth century. Although major parties differ in important ways, they have in recent years combined *classical* liberal appeals to citizens as the bearers of rights and duties and *social* liberal appeals to citizens as productive consumers with appeals to *stakeholder citizens*: self-responsible individuals holding rights to self-orient through the development of personal capabilities and the possession of a 'stake' in society. Political appeals to stakeholders represent the state, businesses, localised communities and individuals as atomistic competitors within an irresistible, juggernaut-like globalising world (Burkitt and Ashton 1996, Kelly, Kelly and Gamble 1997). Its proponents define stakeholder citizenship as 'the ethical and human capital development of the self organised around the possession of stakes' (Prabhakar 2003: 347) within a polity that values competition within a growth-oriented economy as the best way to create and reproduce common goods.

Indeed stakeholder citizenship currently frames engagement with the public by political parties, as well as business and civil society (Scerri 2009).

The rise to prominence of stakeholder citizenship coincides with widespread environmental awareness, often prompted by experience or knowledge of, for example, pollution or extreme weather. Combined with scientific evidence of human-induced climate change, such awareness prompts collective recognition of the need for sustainability. In part stakeholder citizenship reflects the 'eco-anxieties' (Mol and Spaargaren 1993) held by a relatively well-educated and affluent 'post-materialist' public (Inglehardt 1990). In this context, sustainability has emerged as a 'core state interest' (Eckersley 2004). Political parties' 'green' policies and practices are expressed as sustainable development: economic and social development that meets current human needs while not impairing the capacity of future generations to do the same (WCED 1987). However the concept 'sustainability' is imprecise at best. It offers few directives about the actions required to achieve it. Nonetheless, acting to achieve sustainability – by governments, business, civil society and individuals – is increasingly valued as a means of contributing to the common good. At least, efforts to reconceptualise and re-ground values, norms and rules as sustainable are creating a *virtue* out of not supporting degradation of the ecosphere (Connelly 2006).

In this context stakeholder citizenship overplays the importance of individual actions and responsibilities for achieving social goods. This effectively relativises social power and values, undermining collective decision making in favour of individual choices. Stakeholder citizenship tends to subsume claims that the common good is not being achieved under the imperatives of economic growth, making it difficult to claim publicly that growth is itself a central problem for achieving sustainability (Scerri 2009: 476). The 'gospel of eco-efficiency' (Martinez-Alier 2002: 5–10) underpinning sustainable development policy is, in relation to citizenship, premised on a belief that sustainability will be achieved as governments and business manage markets, channelling the trickle down of stakeholder opportunities to develop personal capacities as green consumers and, so, participate in the greening of the economy (Clarke et al. 2007). Aligned with sustainable development policy, stakeholder citizenship provides self-responsible individuals with opportunities to improve their material life conditions and social status in ways that contribute to economic growth and, it is alleged, green outcomes. In the prevalence of stakeholder citizenship, sustainability is represented '*as if* solutions to socially created problems were always synonymous with expedience in the private realm of autonomous sovereign choice' (Scerri 2009: 474).

Sustainable Consumption and 'Discursive Traps'

In this situation stakeholder citizens are called on to contribute to the social good – sustainability – by exercising consumer sovereignty through a privatised ethics of rational choice (Scerri 2006); that is, to assume responsibility for a collective problem by 'ethically' choosing to consume with green discretion (Scerri 2003).

Government, business and some non-government organisations now call on householders to participate proactively in personal use consumption as a way of engaging in sustainability as citizens (Soper 2004). Sustainable consumption is even said by some to give expression to 'democracy through the wallet' (Rayner, Harrison and Irving 2002). Consumers choosing one commodity over another are likened to voters choosing politicians in an election (Schudson 2006). The policy assumption that self-responsible and self-improving citizens will *increasingly* choose to act in virtuous ways is in this sense a pervasive feature of 'green' liberal democracies (Eckersley 2004).

It is in relation to this situation that Hobson demonstrates how policy for sustainable development aims to communicate the 'facts' of climate change to citizens, with the objective of stimulating (green) 'rational' consumption choices. As Hobson shows, sustainable consumption strategies often complement 'eco-efficiency' drives, which promote to householders technological innovations such as energy efficient light globes or water saving showerheads (2006). In these actions, householders' collective rights and responsibilities in relation to the good of sustainability are mobilised and maintained through individuated economic transactions. A key responsibility for stakeholder citizens is to green their own consumption choices while social institutions provide official support for the continual enhancement of personal capacities through the possession of a 'stake' in the greening of local communities.

Kersty Hobson argues convincingly that official, policy-driven interventions to promote household sustainability in Australia and the United Kingdom are largely premised upon achieving just such a rationalisation of individual lifestyles (Hobson 2002). However she also demonstrates that such interventions tend to create *discursive traps*, whereby information supplied to individual householders has the effect of alerting them to and prompting them to think through the nature of structural constraints upon achieving sustainability (Hobson 2003). That is, the phenomenon of the discursive trap countermands the aim of fostering sustainable consumption, prompting citizens to ask, 'Why ... [are we] being told what to do and how to live when governing institutions did little to address [over]consumption issues?' (Hobson 2002: 112). Calling on citizens to 'make consumption changes [is] inextricably bound up [within individuals' worldviews] with broader arguments over social fairness and responsibility' (Hobson 2002: 110).

Using Hobson's concept I now turn to examine empirical evidence in support of my argument that such interventions also bring into question the legitimacy of the compromise between growth and green values. I use the notion of the discursive trap to help make explicit the possibility that a rupture in the legitimacy of this compromise has occurred. By setting Hobson's concept within the analytical schema of the compromise situation, I analyse the questionnaire responses in a way that explains how official policies for achieving the social good of sustainability amongst stakeholder citizens seems to be alerting people to the socio-structural dimensions of the climate change issue. That is, one effect of stakeholder citizenship is to foreground some of what appear to be the *political* distortions and dissimulations around growth as a value orientation. From within the social field of established *cultural* representations that accommodate a compromise between different orders of justification, I use this empirical analysis to argue that the presence of sustainability as a virtue is undermining the pre-eminence of growth in relation to the common good.

Questioning Stakeholder Citizens about Household Sustainability

The questionnaire under consideration formed part of a broader project reporting to a major Australian city council (the City) on the status and uptake of its Residential Sustainability Strategy. The questionnaire aimed to better understand householders' knowledge, trust and beliefs about responsibilities in relation to the City's actions for household sustainability. Policy documents and current practices were analysed on the basis of background documentary and focus group research. This review found the reasons given by the council for taking action – such as making policy, deploying particular initiatives, developing 'indicators' of residential sustainability and for stakeholder engagement – referred to the contribution that householders' participation would make to environmentally sustainable (economic) growth in the city and beyond. By contrast, reasons given by the City for desisting from a course of action, or for denying the reasonableness of a course of action were largely that these were insufficiently participatory and overly authoritative. Importantly, little reference was made in the City's strategy to imbalances in the power of residents in relation to political or economic institutions.

The City also deployed a number of complementary programs, including the Green Tram, Environment Resource Centre and periodical Green Leaflet information programs, as well as the eco-efficient technology-centred Energy Saver—*it all starts at home* – and Showerhead Exchange programs. Another initiative was the localised Sustainability Street program (City of Melbourne 2009). In short, the strategy puts into practice the twin emphases of stakeholder citizenship reliant on individual responsibility and personal capacity building. The City's policy was also in this sense closely aligned with the commitment to sustainability of the Victorian state and Australian federal governments (Department of Infrastructure 2002, Department of Planning and Community Development 2008, Department of Climate Change 2008, Curran 2009). Even though the City's enthusiasm for individual householders' participation in residential sustainability was high, its efforts to establish household sustainability lacked impetus or impact. Indeed, key quantitative indicators chosen by the City to measure residential sustainability, including the Household Carbon Footprint, electricity, water, waste and consumer goods usage rates continue to increase, as do those in all Australian cities (Newton 2007).

The questionnaire contained four true/false and 62 five point 'attitude scale statements'. Between 30 March and 17 April 2009 it was distributed in hard copy to respondents in the City library and in the lobbies of city apartment blocks, as well as by return mail and electronically. The questionnaire also allowed for half-page comments. In total, responses to 25 questions are analysed. The framework for analysis categorised responses in terms of householders' depth of awareness (knowledge) of the sustainability issue; trust in others and social institutions as facilitators of household sustainability and in the City as a source of information on sustainability; and beliefs, values and attitudes in relation to responsibility for sustainability (questions 22, 24, 33, 34, 36–38, 40 based on Nash and Lewis 2006. In total, 343 valid questionnaires were returned, of which 121 supplied open-ended half-page responses. I recognise that within a sampling frame of 71,380 people containing 35,846 residential dwellings (Australian Bureau of Statistics 2006), 343 responses should be regarded cautiously as a representative sample of this population.

Knowledge

True/false questions in this section included:

33.   Current rates of resource use can be maintained indefinitely.
34.   A sustainable future is only possible if levels of material wealth are lowered.
36.   Local species of plants and animals are disappearing rapidly because of human impacts.
47.   I know enough about global warming.
48.   On an average day, I do not do anything personally that causes greenhouse gas emissions.
49.   Restrictions on household water use currently apply in the city.
50.   The hole in the ozone layer is the main cause of global warming.
51.   Plastic bags should go into household recycling.
52.   Air-conditioning doesn't contribute to household energy use.
44.   I feel overwhelmed by information on climate change.
64.   It is often difficult to act in environmentally friendly ways because I don't know where to begin.
65.   I try and buy less stuff to reduce the environmental impact of my lifestyle.

Responses in this category demonstrated that the sample population was quite knowledgeable about the 'facts'. Understanding of the long-term issues surrounding current rates of human exploitation of the ecosphere was demonstrated by the 83 per cent of respondents who strongly disagreed or disagreed with Question 33: 'Current rates of resource use can be maintained indefinitely.' Similarly 55 per cent of respondents agreed or strongly agreed with Question 34: 'A sustainable future is only possible if levels of material wealth are lowered.' This suggests a high level of agreement with the view that current material standards of living are unsustainable. Moreover these responses imply that householders understand that (economic) growth and sustainability (of society within the ecosphere) are incompatible. A small majority believed that local species of plants and animals are disappearing rapidly due to human impacts (79 per cent of respondents strongly agreed or agreed). This is interesting insofar as inner Melbourne is one of the most densely populated and long-established urban areas in Australia, containing few if no areas of native bushland. This response demonstrates householders' appreciation of the holistic nature of the sustainability issue: it is a holistic virtue, even when the focus is expressly upon residential sustainability in the city's core.

An insignificant majority disagreed or strongly disagreed with Question 47: 'I know enough about global warming.' This may suggest a need to communicate more 'factual' information to householders, but also hints at fatigue with the issue. This is reinforced by responses to Question 44: 'I feel overwhelmed by information on climate change' and Question 64: 'It is often difficult to act in environmentally friendly ways because I don't know where to begin.' Responses to Question 44 were most commonly 'agree'; with 53 per cent disagreeing or strongly disagreeing with Question 64. This suggests a disjuncture between knowing what to do and taking action: most householders have some idea of 'where to begin' but many feel overwhelmed by the 'facts'. One respondent expressed her/his regret at not having the economic means to 'do more': 'I do recycle and compost, have [a] dual flush loo and water reducing shower hose but cannot afford solar panels or water tanks although I agree with them.' Another commented, 'There is only so much that the majority of individuals will want to do. If the problem is to be tackled, the focus needs to be on large-scale reductions from the big [industrial] consumers'.

Although 71 per cent of respondents disagreed with Question 48: 'On an average day, I do not do anything personally that causes greenhouse gas emissions', a significant proportion (16 per cent), agreed or strongly agreed. The most common response was 'disagree', suggesting recognition amongst householders that *all* social activities generate greenhouse gases. Such responses give an objective picture of a situation in which a significant minority either fail to understand that urban households always cause greenhouse emissions or, more importantly, feel that their 'offsetting' efforts actually *negate* individual or collective contributions to climate change (Bailey and Wilson 2009, Smith 2007). Indeed three open-ended responses indicated directly that householders felt this was the effect of 'offsetting' activities. Such responses can in part be understood in terms of widely publicised commercially driven and state-sponsored attempts to justify 'carbon offsetting' and carbon trading schemes, even though significant research has shown that such policies will not have the effect of reducing carbon emissions or will not reduce them in time to avert problematic consequences (Lohmann 2006, Takahashi 2004, Greenspan-Bell 2006, Harriss-White 2008).

Questions 49–52 were formatted as true/false/unsure/no opinion questions. Only 17 per cent of respondents were unsure or reported false in relation to water restrictions then applying in the city, while 95 per cent agreed that domestic air-conditioning contributes to household energy use. While these figures are encouraging, the statement as 'scientific fact' that the hole in the ozone layer is the main cause of global warming was answered 'true' or 'unsure' by 40 per cent of respondents, indicating a poor understanding of the atmospheric processes associated with global warming. The statement that 'Plastic bags should go into household recycling' was answered 'true' or 'unsure' by 46 per cent of respondents. The relatively large percentage of respondents who affirmed or were unsure about this (false) statement again suggests that recycling drives have on the whole the effect of misleading people into thinking that all plastics are recyclable. It might even be reasonable to imply that recycling discourses undermine those of reuse and, more importantly, those of 'don't use in the first place'. Responses to Question 65: 'I try and buy less stuff to reduce the environmental impact of my lifestyle', emphasised householders' beliefs in links between overconsumption and unsustainability: 63 per cent agreed or strongly agreed with the statement. The presence of a significant minority (28 per cent) who disagreed or disagreed strongly with this statement (5 per cent had no opinion and 4 per cent reported neither) may demonstrate a marked degree of recognition that lowering household consumption can only go so far in addressing the sustainability issue.

Trust

Questions in this section included:

11. I feel that formally-educated experts, such as government administrators, scientific experts and managers, can be trusted when dealing with local issues.
12. I feel that governments make decisions and laws that are good for the way I live locally.
13. I feel that most people can be trusted.
22. Experts will always find a way to solve environmental problems.
24. Conserving natural resources is unnecessary, because alternatives will always be found.
25. Industry is doing everything it can to prevent environmental damage.
37. The most important concern of residents should be the city's economic growth.
43. I trust City information on climate change.

Responses to Question 11: 'I feel that formally-educated experts, such as government administrators, scientific experts and managers, can be trusted when dealing with local issues', suggest that there is a high level of trust of institutionalised expertise amongst residents (38 per cent agreed or strongly agreed). This complements responses to Question 12: 'I feel that governments make decisions and laws that are good for the way I live locally', which were broadly comparable. Significant proportions of the sample (59 per cent) agreed or strongly agreed with Question 13: 'I feel that most people can be trusted.' Responses to Question 43: 'I trust City information on climate change', represented a significant degree of ambivalence (41 per cent responding 'neither', which was also the most common response).

The responses to Question 11 stand in contrast with responses to Question 22: 'Experts will always find a way to solve environmental problems' (45 per cent disagreed or strongly disagreed), which demonstrates a marked pessimism about the overall capacity of social institutions to meet the challenge of sustainability through the application of expert knowledge or practice informed by experts. These responses suggest that the sheer complexity of scientific knowledge associated with the sustainability issue, combined with the questionable status of information from commercially motivated information campaigns, often raises doubts within the citizenry about householders' ability to effect substantive change, yet householders still rely on experts for leadership. That is, while householders believe that 'formally-educated experts, such as government administrators, scientific experts and managers, can be trusted', levels of trust in current institutional arrangements that privilege experts is significantly lower. Respondents were deeply opposed to Question 37: 'The most important concern of residents should be the city's economic growth', with 62 per cent disagreeing or strongly disagreeing with this statement. This was supported by responses to question 33 (see above) and, more emphatically, by responses to Question 25: 'Industry is doing everything it can to prevent environmental damage', which was overwhelmingly disagreed or strongly disagreed with by 83 per cent of respondents. Similarly 86 per cent disagreed or strongly disagreed with Question 24: 'Conserving natural resources is unnecessary, because alternatives will always be found.'

These responses suggest that householders construe as ineffective policies to engender household sustainability which rely on individual or business voluntarism. I want to propose that this situation actually weakens public trust in the 'truthfulness' of information and communication strategies and the effectiveness of institutional actions. These responses reinforce the point that, while not completely conversant in the abstractions of scientific 'facts', householders nonetheless understand that social life *as it is currently organised* is unsustainable. These responses suggest the presence of a significant effectiveness barrier for institutions that orient sustainability strategies in a singular way towards communication and individual householder participation. That is, these responses demonstrate what happens at the interstices of what Hobson calls discursive traps, as well as a marked ambivalence towards the legitimacy of current institutional approaches to managing the transition to household sustainability. Moreover, and although further discussion of this issue lies beyond this one, these responses imply that householders' distrust of the capacity of social institutions, as currently organised, to achieve sustainability is not necessarily related to widespread rejection in general of social institutions based on 'radical' environmental Anarchist or Romantic values.

Responsibility

Questions in this section included:

38.   As residents, we have a moral obligation to protect our local areas of green space.

39.   City residents need to be more environmentally aware in the future.

40.   Urban expansion must be stopped to protect remaining green areas.

42.   Reducing the environmental impact of every day life is the responsibility of every resident.

45.   City information about sustainability is a valuable contribution to Australia's attempts to address climate change.

46.   The City provides enough practical support to residents who want to reduce the environmental impact of their lifestyles.

53.   Household environmental sustainability should be the responsibility of the resident.

54.   It is appropriate for the City to link the cost of rates to environmental features of a property (e.g. insulation, water tanks etc).

55.   It is appropriate for the State and/or Federal government to legislate certain environmental features of residential properties.

These questions were sub-categorised in terms of 'individual' or 'institutional' responsibility. Responses across Questions 38, 39, 42 and 53 consistently represented a high level of agreement among householders that residential sustainability supplies a legitimate moral imperative that individuals should act on (on average 72 per cent agreed or strongly agreed with these statements). However responses to Question 54: 'It is appropriate for the City to link the cost of rates to environmental features of a property (e.g. insulation, water tanks etc.)' and Question 55: 'It is appropriate for the State and/or Federal government to legislate certain environmental features of residential properties' were similarly emphatic. In relation to Question 54, 59 per cent agreed or strongly agreed, while 80 per cent agreed or strongly agreed that 'it is appropriate for State and/or Federal governments to legislate certain environmental features of residential properties'. These responses suggest that a reaction to the privileging of the individual and voluntary over collective and regulatory responsibility for achieving residential sustainability is emerging. This is reinforced by responses to Question 55 (80 per cent agreed or strongly agreed) which taken together suggest that institutional action is highly valued and, indeed, is being actively called for by householders. This is reinforced by responses to Question 40, where 51 per cent agreed or strongly agreed. That is, respondents were willing to countenance and support increased constraint on social actors as well as on growth-oriented policy making directly (see also responses to Questions 33 and 34 above).

It is important to note that individual moral responsibility does not exclude a strong belief that the City, as well as state and federal governments are responsible for taking action to achieve sustainability. This holds, to a certain extent, even if such actions raise the economic costs of sustainability or reduce possible consumer opportunities for householders. Understood in this context, responses to Question 45: 'City information about sustainability is a valuable contribution to Australia's attempts to address climate change', with 59 per cent agreeing or strongly agreeing, suggest that communications strategies are seen as merely complementing an overall collective response to residential sustainability. Indeed the majority of open-ended responses expressed consternation at the low level of governmental regulation on the household sustainability issue and on climate change more generally. These included succinct statements such as, 'I look forward to stronger council and state government support/direction in relation to sustainability' and longer, elaborate prescriptions for what each tier of government should do – including adjusting rates, legislating that developers include water and electricity saving devices, supplying bicycle lanes on major roads, subsidising solar and wind energy, and harsher penalties for ecological miscreants – as well as calls for the removal of 'arcane rules about [air] drying clothes on [apartment] balconies', which force householders to use electric dryers.

Conclusion

Throughout this chapter I have applied a critical pragmatic analytical framework to combine theoretical claims about the prevalence of a particular form of citizenship with empirical research based on a questionnaire aimed at understanding beliefs and values in relation to household sustainability. My analysis exposes a nascent rejection of the privileging of 'rational' consumption choices and institutional voluntarism in official strategies for increasing household sustainability. This analysis supports my assertion that sustainability, as a virtue, is supporting a shift in values away from an unstable 'compromise' between an order of justification that privileges growth and one that defines the common good in terms of creating and maintaining a sustainable relationship between society and the ecosphere. Implicit in my argument is the claim that among householders 'green' values are taking precedence over those based in 'growth'.

As citizens, householders recognise that sustainability is a virtue and as such it entails private moral obligations while also justifying collective action. My analysis demonstrates that householders are aware of the fallibility of individual understandings of 'the facts', and recognise that individual and institutional voluntarism is failing to achieve sustainability, whether in relation to households or more generally. Indeed, responses to the questionnaire suggest that sustainability and household sustainability are practically inseparable. Householders see 'greenness' itself as a way of defining the common social good, and that such a greening of values is in some ways opposed to growth. These findings concur with similar research in Australia and the United Kingdom (for example see Burgess, Harrison and Filius 1998, Hobson 2004, Flynn, Bellaby and Ricci 2008, Shepherd 2009, Jorgensen, Graymore and O'Toole 2009). Householders seem well aware of the limitations of their own knowledge of 'the facts', and are explicit in believing that government and business should ensure that the responsibility to act sustainably is held collectively.

A key implication of these findings is that although largely problematic as material contributions to household sustainability, sustainable consumption and eco-efficient strategies for achieving household sustainability can be re-conceptualised as tools for promoting value change and knowledge about the social relations that support unsustainability. The research report presented to the City recommended that its household sustainability strategy would be more efficacious if it (1) provided a forum for asserting the integrity of scientific 'facts', (2) acted as an advocate for 'green' identity by explaining the social relations behind consumption choices (3) was a platform for advocating greater local government involvement in developing and implementing state and federal sustainability policy.

Given that lifestyle changes made by individuals may be of limited value in the broader context of a growth-oriented economy (Wapner and Willoughby 2005: 84), and the householders surveyed seem to recognise this, there is a need to alert householders to the ways in which local government is addressing the structural conditions and social relations that currently impede actions to achieve sustainability. In this view, sub-national institutions, such as city councils, may more effectively use their status and economic resources to advocate and promote value change that aims to foster a green 'identity' (Crompton and Kasser 2009). This would entail a shift away from communicating scientific 'facts' and eco-efficient 'tips'. It would encourage policy makers to develop information that explains how and why consumption choices within a growth-oriented economy are not always automatically ecologically beneficial, with a view to encouraging a broader perspective on ecologically relevant actions at the household scale. Aiming the strategy in these directions, it was suggested, would benefit the City by: (1) situating the City as a (possibly leading) source of information about the reliability of scientific data and the integrity of science on the urban household sustainability issue, (2) increasing its public profile as an advocate for household sustainability as a socially constituted problem, and (3) becoming an important node within the growing global and national networks of sub-national actors that are influencing policy for sustainable development.

While this analysis supports my own normative preference for policy and planning that regulates individual and institutional actions in relation to environmental issues, it also supports my analytical claim that the 'compromise' central to policy interventions directed at stakeholder citizens is breaking down. Building on Hobson's work, my analysis supports a prediction that the maintenance and expansion of stakeholder-oriented interventions into householders' material practices will have the effect of further undermining the legitimacy of such policy. Whether or not this unease with voluntarism and self-regulation translates into political support for greenness, over growth, remains to be seen. It may indeed undermine public support for sustainability itself by fostering disillusionment with the justice of the distribution of responsibility among individuals, business and the state. Nonetheless I believe that this shift has implications for both sustainable development and stakeholder citizenship. In the macro social terms of sustainable development, my analysis suggests the presence of public support for stronger regulatory approaches by government to ensure the just distribution of individual and institutional responsibility. In the micro social terms of stakeholder citizenship, my findings imply support by citizens for greater government intervention to reduce levels of material consumption and resource use. One respondent summed up feelings of exasperation with the privileging of individual responsibility and personal capacity: 'I feel that the most effective way to improve household sustainability is for State and Federal governments to LEGISLATE for change.'

References

Australian Bureau of Statistics 2006. *Census Data* [Online]. Available at: www. censusdata.abs.gov.au/ABSNavigation/prenav/LocationSearch?collection=Ce nsus&period=2006&areacode=LGA24600&producttype=QuickStats&breadc rumb=PL&action=401 [accessed: 8 September 2009].

Bailey, I. and Wilson, G.A. 2009. Theorising transitional pathways in response to climate change: technocentrism, ecocentrism, and the carbon economy. *Environment & Planning A*, 41(10): 2324–2341.

Boltanski, L. and Thévenot, L. 1999. The Sociology of Critical Capacity. *European Journal of Social Theory* 2(3): 359–377.

Boltanski, L. and Thévenot, L. 2006 [1991]. *On Justification: Economies of Worth*. Translated by C. Porter. Princeton: Princeton University Press.

Burgess, J., Harrison, C.M. and Filius, P. 1998. Environmental communication and the cultural politics of citizenship. *Environment and Planning A*, 30(8): 1445–1460.

Burkitt, B. and Ashton, F. 1996. The birth of the stakeholder society. *Critical Social Policy*, 16(49), 3–16.

Chiapello, E. 2003. Reconciling the two principal meanings of the notion of ideology: the example of the concept of the 'new spirit of capitalism'. *European Journal of Social Theory*, 6(2): 155–171.

City of Melbourne 2009. Home Page [Online]. Available at: www.melbourne.vic. gov.au/info.cfm?top=218&pg=4017 [accessed 11 June 2009].

Clarke, N., Barnett, C., Cloke, P. and Malpass, A. 2007. Globalising the consumer: doing politics in an ethical register. *Political Geography*, 26(3), 231–249.

Connelly, J. 2006. The virtues of environmental citizenship, in *Environmental Citizenship*, edited by A. Dobson and D. Bell. Cambridge, MA: MIT Press, 49–74.

Crompton, T. and Kasser, T. 2009. *Meeting Environmental Challenges: The Role of Human Identity*. Godalming, UK: World Wildlife Fund UK and Green Books.

Curran, G. 2009. Ecological modernisation and climate change in Australia. *Environmental Politics*, 18(2): 201–217.

Daly, H. 1999. *Ecological Economics and the Ecology of Economics*. Cheltenham: Edward Elgar.

Department of Climate Change. 2008. *The Carbon Pollution Reduction Scheme: Green Paper*. Canberra: Department of Climate Change.

Department of Infrastructure. 2002. *Melbourne 2030: Planning for Sustainable Growth*. Melbourne: Department of Infrastructure.

Department of Planning and Community Development. 2008. *Melbourne 2030: A Planning Update – Melbourne @ 5 million*. Melbourne: Department of Planning and Community Development.

Dumont, L. 1986 [1983]. *Essays on Individualism: Modern Ideology in Anthropological Perspective*. English edition. Chicago: University of Chicago Press.

Eckersley, R. 2004. *The Green State: Rethinking Democracy and Sovereignty*. Cambridge, MA: MIT Press.

Flynn, R., Bellaby, P. and Ricci, M. 2008. Environmental citizenship and public attitudes to hydrogen energy technologies. *Environmental Politics*, 17(5): 766–783.

Greenspan-Bell, R. 2006. Kyoto placebo. *Issues in Science and Technology*, Winter: 28–31.

Harriss-White, B. 2008. Market politics and climate change. *Development*, 51(3): 350–358.

Hobson, K. 2002. Competing discourses of sustainable consumption: does the 'rationalisation of lifestyles' make sense? *Environmental Politics*, 11(2): 95–120.

Hobson, K. 2003. Thinking Habits into Action: the role of knowledge and process in questioning household consumption practices. *Local Environment*, 8(1): 95–112.

Hobson, K. 2004. Sustainable consumption in the United Kingdom: the 'responsible' consumer and government at 'arm's length'. *Environment & Development*, 13(2): 121–139.

Hobson, K. 2006. Bins, bulbs and shower timers: on the 'techno-ethics' of sustainable living. *Ethics, Place & Environment*, 9(3): 317–336.

Hutchinson, F., Mellor, M. and Olsen, W. 2002. *The Politics of Money: Towards Sustainability and Economic Democracy*. London: Pluto.

Inglehardt, R. 1990. *Culture Shift in Advanced Industrial Societies*. Princeton: Princeton University Press.

Jorgensen, B., Graymore, M. and O'Toole, K. 2009. Household water use behavior: an integrated model. *Journal of Environmental Management*, 91(1): 227–236.

Karagiannis, N. and Wagner, P. 2008. Varieties of agonism: conflict, the common good, and the need for synagonism. *Journal of Social Philosophy*, 39(3): 323–339.

Kelly, G., Kelly, D. and Gamble, A. 1997. *Stakeholder Capitalism*. London: Macmillan.

Lohmann, L. 2006. Carbon trading: a critical conversation on climate change, privatisation and power. Special edition, *Development*, 48: 1–359.

Marshall, T.H. 1998 [1950]. Citizenship and social class, in *The Citizenship Debates: A Reader*, edited by G. Shafir. Minneapolis: Univeristy of Minnesota Press, 93–111.

Martinez-Alier, J. 2002. *The Environmentalism of the Poor: A Study of Ecological Conflicts and Valuation*. Cheltenham: Edward Elgar.

Mol, A. and Spaargaren, G. 1993. Environment, modernity and the risk-society. *International Sociology*, 8(4): 431–459.

Nash, N. and A. Lewis. 2006. Overcoming Obstacles to Ecological Citizenship: The Dominant Social Paradigm and Local Environmentalism. In Environmental Citizenship, edited by A. Dobson and D. Bell. Cambridge: MIT Press, 2006, 153-84.

Newton, P. 2007. Horizon 3 planning: meshing liveability with sustainability. *Environment and Planning B*, 34(4): 571–575.

O'Connor, J. 1996. The second contradiction of capitalism, in *The Greening of Marxism*, edited by T. Benton. New York: Guilford Press, 197–221.

Perelman, M. 2003. *The Perverse Economy: The Impacts of Markets on People and the Environment*. New York: Palgrave.

Prabhakar, R. 2003. Stakeholding: does it possess a stable core? *Journal of Political Ideologies*, 8(3): 347–363.

Rayner, M., Harrison, R. and Irving, S. 2002. Ethical consumerism: democracy through the wallet. *Journal of Research for Consumers*, 1(3): 1–12.

Scerri, A. 2003. Triple bottom-line capitalism and the 3rd Place. *Arena Journal*, New Series 20: 56–67.

Scerri, A. 2006. Self-Orienting Individuals: Subjectivity and Contemporary Liberal Individualism. Thesis submitted in partial fulfilment of PhD award, Melbourne, RMIT University.

Scerri, A. 2009. Paradoxes of increased individuation and public awareness of environmental issues. *Environmental Politics*, 18(4): 467–485.

Schudson, M. 2006. The troubling equivalence of citizen and consumer. *Annals of the American Academy of Political and Social Sciences*, 608: 193–204.

Shepherd, J. 2009. *Geoengineering the Climate: Science, Governance and Uncertainty*. London: The Royal Society.

Smith, K. 2007. *The Carbon Neutral Myth: Offset Indulgences for Your Climate Sins*. [Online: Carbon Trade Watch]. Available at: www.carbontradewatch. org/index.php?option=com_content&Itemid=327&id=288&task=view< [accessed: 14 May 2009].

Soper, K. 2004. Rethinking the 'good life': the consumer as citizen. *Capitalism, Nature, Socialism*, 15(3): 111–116.

Takahashi, T. 2004. The fate of industrial carbon dioxide. *Science*, 305(5682): 352–353.

Thévenot, L., Moody, M. and Lafaye, C. 2001. Forms of valuing nature: arguments and modes of justification in French and American environmental disputes, in *Rethinking Comparative Cultural Sociology*, edited by L. Thévenot and M. Lamont. Cambridge: Cambridge University Press, 229–272.

Valdivielso, J. 2005. Social citizenship and the environment. *Environmental Politics*, 14(2): 239–254.

Wapner, P. and Willoughby, J. 2005. The irony of environmentalism: the ecological futility but political necessity of lifestyle change. *Ethics & International Affairs*, 19(3): 77–89.

WCED, *see* World Commission on Environment and Development

World Commission on Environment and Development. 1987. *World Commission on Environment and Development: Our Common Future*. Oxford: Oxford University Press.

Chapter 11

Environmental Politics, Green Governmentality and the Possibility of a 'Creative Grammar' for Domestic Sustainable Consumption

Kersty Hobson

Introduction

> [I]t is necessary to take on board the complexity of the social practices involved in shifting toward more sustainable lifestyles, mediated through systems of provision, if we are to develop a more realistic view of the trajectories and possibilities of future change. In this respect, whilst we can be positive about the grassroots commitments that are emerging around the climate-change agenda ... policy communities need to recognise that far more than a shift in the attitudes and intentions of individuals is required to achieve significant carbon reductions through these means. (Walker and Cass 2007: 467)

I open this chapter with the above quote as it offers a salient echo of arguments social and cultural researchers have made over the past few decades around the substantial challenges faced in moving current resource intensive domestic practices towards more sustainable trajectories (for example Burgess et al. 2003). Indeed Gordon Walker and Noel Cass's above assertion speaks to a now substantial body of quantitative and qualitative research into the often bemusing 'complexity of social practices' (Walker and Cass 2007) that undergirds modern lifestyles: research that has, for one thing, put paid to any notion that fostering domestic sustainable consumption practices can be achieved via piecemeal education campaigns or financial incentives (for example Geller 1992). Instead, this body of research has underscored how habits, unsustainable infrastructures, norms, personal idiosyncrasies, multiple demands on all our time and energy, and many other factors have confounded attempts by multiple governmental actors to get everyone 'doing their bit' for the environment. As a result – and despite manifold exhortations to do otherwise – domestic resource use continues to climb in countries all around the world, 'developed' or otherwise (for example Worldwatch Institute 2008).

Thus as Walker and Cass point out, much more is indeed needed than just shifts in purported public attitudes if the burgeoning challenges of sustainability are to be met. However what the 'far more' that Walker and Cass allude to actually entails is the subject of ongoing debate. For some it means less focus on the small and individualised actions of, for example, recycling household waste or riding a bike to work. Instead the emphasis of individuals and non-government organisations in particular should be placed on mobilising collective movements of sustainable practices and political pressure (for example Maniates 2002); although the plausibility of such a shift is itself contentious, given debates about the apparent decline in civil society, evidenced by the seeming unwillingness of citizens to partake in many forms of collective action (for example Putnam 2000, Macnaghten 2003). For others the 'ecological modernisation' of the domestic sphere via highly efficient technologies and infrastructure is key, where '[s]ustainabilty is, then, in part, delegated to a material landscape in which the non-human actors (machines, devices, infrastructures) translate the actions of the human inhabitants automatically towards eco-friendly outcomes' (Jelsma 2003: 104).

Of course many other approaches to promoting domestic sustainable consumption exist and no single agenda can or should prevail. Rather a multi-faceted toolbox for stimulating household behaviour change is logical and necessary, and indeed typifies the plethora of current approaches that include information campaigns and fiscal incentives.[1]

The intention of this chapter however is not to survey the contents of this policy and practice toolbox per se. Rather it is to explore some assumptions that I argue underpin various approaches to encouraging lifestyle changes: assumptions that have potentially significant impacts on framings and outcomes but often remain under-interrogated in the literature. More specifically, in this chapter my opening question simply put is: even if millions of people 'doing their bit' through small practice changes has yet to accumulate into a world-saving state of sustainable consumption, do the everyday acts of recycling and altering household fixtures not prime us all to become more aware and active 'environmental citizens' in the long run, making it more likely that we will take further pro-environmental steps in the future (for example see Verplanken and Holland 2002)? Or does this form of sustainable consumption represent a glass ceiling (or rather, glass picket fence) that constrains and limits the scope of possible engagements with the challenges of sustainability past those that are low or no-cost, small-scale and 'stuff-focused', resulting in an exaggerated sense of the contributions that 'our bit' of reusing shopping bags and composting makes to overall sustainability (Whitmarsh 2009)?

---

1   For example the New South Wales Department of Environment, Climate Change and Water has a 'Home Savers Rebate' scheme for the fitting of, for example, domestic rainwater tanks and dual flush toilets (see http://www.environment.nsw.gov.au/rebates). It also provides information on altering various domestic practices, such as water use, household cleaning and recycling (see http://www.environment.nsw.gov.au/households/index.htm).

Thus this chapter takes up these questions and aims to reflect on their premises through recent work on materialities of household sustainable consumption. This particular lens is adopted here as in the past inadequate attention has been paid to the vital role that material goods and their associated embodied practices play in questions of sustainable lifestyles, although this has recently started to change (for example Cupples and Ridley 2008). In this chapter I discuss findings from my own research into the attempts of UK and Australian households to reduce their everyday resource consumption. I outline how the materialities of domestic sustainable consumption do indeed play a crucial, but ambivalent, role in sustainable lifestyle behaviour change. I argue that they are instrumental in enacting particular forms of the modern 'environmental subject' in keeping with prevailing modes of governmental intervention, discussed here through the tenets of Foucault's writing on governmentality. That is, a subject who achieves a form of sustainability via the acquisition of particular 'green goods' and who, in the process (purposefully, at times), side-steps imperatives to participate in a form of environmental politics focused on issues and action outside the home. However rather than conclude that a materialised individualised form of sustainable consumption heralds the demise of environmental politics per se, I reconsider here some key concepts at play – such as what constitutes environmental politics and political action in this context – and go on to suggest that such a trajectory might benefit from a more 'creative grammar' in the forging of or and engagements with sustainable material cultures.

Forging the Environmental Citizen: On Multiple Interventions and Green Governmentality

Why Does Domestic Unsustainability Persist?

Of late, policy interventions into the realm of sustainable lifestyles and domestic behaviour change have proliferated in form. For example fiscal incentives for altering home energy infrastructures are now commonplace (see above), and the experimental testing of energy systems feedback has shown promising if not somewhat inconclusive results (for example Wood and Newborough 2003). In addition widespread information campaigns in various guises have informed households about what they can do to 'make a difference'. These have taken many forms, including one-off voluntary interventions such as online carbon calculators, which sum up and advise on ways to reduce one's 'ecological footprint' (for example http://www.abc.net.au/tv/carboncops/calculator.htm). There have also been more in-depth and sustained programs that take participants through some of the step-by-step processes involved in changing a whole suite of resource intensive practices (for example see Staats and Harland 1995, Hobson 2001).

While not without impacts and positive effects (for example Hobson 2003) researchers from across the social and human sciences have underlined some key problematic issues in such approaches. For example it has been argued that simply giving individuals information about what they can/should/could alter in their everyday routines and material landscapes of the home is an inadequate mechanism for instigating noticeable and sustained domestic behavioural changes. Why this is the case is argued through multiple conceptual and methodological framings that together present an intricate and at times seemingly inconsistent picture of human intentionality and interactions with the materialities of the home. For example there is the 'rebound effect', wherein greater technological efficiency is outstripped by absolute increases in the use of a particular technology (Science and Technology Committee and The House of Lords UK 2005). There are also indirect and fragile links between expressed individual environmental concerns and particular behavioural outcomes (Whitmarsh 2009); disconnect between abstract entities like 'the environment' and 'climate change', and everyday lives and concerns (Macnaghten 2003); and the unsustainable, deeply entrenched and ponderous nature of prevailing socio-technical systems of domestic provision (Shove 2003).

As such, any pretence that quick-fix, informational or fiscal interventions can address the increasing rise of unsustainable domestic consumption has been thoroughly challenged. So if this is the case – and if indeed the complexity of our everyday practices appears to overwhelm attempts to make society more sustainable – what are we to do? Without suggesting there is a simple response that can fully satisfy this inquiry, in the remainder of this chapter I explore this question further, beginning by drawing on the work of human geographers and other cultural-social researchers into the governmental intentions of domestic sustainable consumption interventions.

On the Advent of Green Governmentality

From a pragmatic perspective attempts to make us all 'do our bit' at home obviously aim to contribute to the pressing need to reduce global greenhouse gas emissions. However as the Australian socio-economic analyst and writer Ross Gittins (2009: 1) has argued, 'nothing we choose to do for moral reasons will do anything to reduce the nation's total emissions of greenhouse gases' if a national emissions cap means any savings in the domestic sphere allows other sectors to pollute more. If this is the case, the question needs to be asked why Australians are being told that 'Everyday choices can make a difference: from how you travel, to saving energy, to your shopping choices' (Department of Climate Change and Energy Efficiency 2010). On the one hand this statement is made because it is indeed true: our choices *can* make a difference, depending on national policy frameworks such as the management of emission caps across various sectors. But on the other hand researchers exploring such questions have pointed out that there are political rationalities at work here that require further consideration (for example Hobson 2004). Indeed of late, politically inflected theories like Foucault's writings on governmentality have been taken up by social researchers from across the disciplinary spectrum to explore the nature of the political subjects being 'worked up' by such forms of intervention (see Rutherford 2007).

To put this work in a broader intellectual context, research into modern forms of governance has underscored how the purportedly private nature of the practices, relationships and decisions of the home are in fact far from it. That is, these aspects of our lives are, rather, subject to governing forces commonly framed as existing in the realm of the political, regulatory and thus 'public' domain (see Staeheli and Mitchell 2004). Whether it is planning laws around the installation of domestic water tanks (for example see Troy 2008), or incentives to encourage higher births per capita,[2] such decisions are often subject to exogenous moral, financial or regulatory suasion in various guises.

Michel Foucault's work on governmentality has been a highly productive lens through which researchers have explored such intersections between the private and public realms: that is, the manifold, diffuse and often contested means by which neo-liberal forms of governance operate. Working up his ambitious 'genealogy of the state' before his death in the mid-1980s, Foucault's aim – in-keeping with his long-held views on the ongoing and partial nature of inquiry and knowledge (for example Foucault 2004) – was not to arrive at a picture of or definitive theory about what the state is. Rather, he was interested in how modern polities function. That is, how specific governance rationales and aims are brought into being and enacted through diverse mechanisms that attempt to enrol a wide array of actors into meeting specific ends. Here the 'old ways' of governance still persist, as the state retains many of its traditional governing functions through direct interventions. But in addition new and diverse techniques of intervention have emerged. As a result: 'Governance is being reconfigured from the level of the nation-state to the local community through technologies of power that seek to promote active agency, responsible self-governance and the state as an enabler as opposed to provider of services' (McKee 2008: 184). Foucault's original ideas were focused in particular on exploring the centrality of discourses to modern modes of governing. Here he asserted that the discourses traded in and deployed to communicate and understand the world do not just describe life 'as it is'. Rather they function to constitute it at its most fundamental level: to designate that which can and cannot be said, thus emphasising the 'awesome materiality' of language (Foucault cited in Slocum 2004: 765).

Scholars taking up Foucault's ideas have thus focused on many aspects of modern governance, all of which obviously cannot be surveyed here (see for

---

2   In 2004 the Treasurer in Australia's right-wing Coalition government's, Peter Costello, encouraged Australian residents to counter the nation's ageing population profile and declining birthrate by having three babies per couple. As he infamously proclaimed, 'have one for mum, one for dad, and one for the country'. Tax breaks and various other payments such as the 'Baby Bonus' were made available to new parents: a move that this now ex-government later claimed was responsible for Australia's apparently increasing birth rate (see BBC 2006).

example Huxley 2008). Those most relevant to this chapter include explorations of governing techniques that work to constitute political subjectivities as desired by the various components of the state and a broad array of policy actors; such techniques as fostering subjects to embody civic responsibilities through notions of 'ecological', 'environmental' or 'energy' citizenship (for example Luque 2005, Rutherford 2007). Such work has led some to conclude that the discourses and interventions outlined above around domestic sustainable consumption aim to 'work up' and bring into existence '[a] responsible, carbon-calculating individual ... the vision of a self-reflexive individual taking responsibility for knowing and reducing his or her emissions' (Rutland and Aylett 2008: 642).

While such arguments are useful in placing the domestic sustainable consumption agenda in broader socio-political contexts there are two important points to be made here. Firstly, governmentality scholars have also argued that the intentions of governmental interventions are not seamlessly translated into desired outcomes. That is, there is always room for 'slippage' and resistance by subjects (Raco 2003). Secondly, and closely related to the first point, the post-structuralists' emphasis on discourses and their performances has paid insufficient attention to the physicalities and materialities of the enactments of discourses, creating a picture of the individualised and placeless subjects: 'one abstracted from personal lived history, as well as from historical and geographical embeddedness' (Nelson 1999: 332).

Thus the impetus now exists in governmentality studies to pay greater attention to the technologies, artefacts and material landscapes that are more than just the setting of governmental interventions. Indeed these materialities are deemed essential to marking the boundaries of that which is possible or not via such interventions (see Merriman 2005). That is, if sustainable lifestyle interventions from national governments or other policy actors aim to create a 'responsible, carbon-calculating individual' (Slocum 2004: 765), there is a definite need to enquire about the role that the materialities of such interventions play in their outcomes.

In response, in the next section, I explore this point further, drawing on empirical research into domestic sustainable consumption programs in Australia, and to a lesser extent the United Kingdom, to argue that indeed the 'stuff' of 'green living' plays a crucial role in forging and re-embedding particular forms of the environmental citizen-subject: one with a tendency to eschew collective and politically focused forms of social change action that some commentators favour as a preferred mode of modern environmental politics.

The Ambivalence of 'Green Stuff'

Over the past decade I have undertaken qualitative research into domestic sustainable consumption programs in both the United Kingdom and Australia. My research focus has been on the experiences of program participants trying to alter their everyday practices, as well as the policy-political contexts of which such programs are a part (Hobson 2002, 2006a, 2006b, 2006c). This section draws on findings from this work: in particular the Australian case study, as the importance of materialities in domestic sustainable consumption practices became most noticeable during research with individuals taking part in the Australian Conservation Foundation's GreenHomes project (see Hobson 2006c).

Although this work did not start out intending to examine the materialities of sustainable consumption, it became increasingly apparent during interviews and participant observation fieldwork that the 'stuff' of household sustainability has a presence in questions of sustainable consumption that had previously been under-acknowledged. This recognition resonated with the recent 'materialist turn' in social science disciplines such as human geography, which provides theoretical arguments for the role of materialities as actants in everyday lives to be taken seriously. This work questions simple understandings of matter or materiality that attribute fixed properties to the material landscape (Anderson and Tolia-Kelly 2004: 669). For example Kaika's (2004) 'nature/cultures' writing argues that the space of the home has constituted entities like water as hybrid or quasi-objects; that is, objects that are materially produced as a commodity through pipes and water bills while also being socially constructed as part of nature – making them both 'in here' and 'out there' at the same time. As such, water does not have a fixed identity or invoke specific practices in itself; rather, it is part of an inconstant and protean entity.

In the case of the GreenHomes research participants, there was certainly evidence that the materialities that have become part and parcel of moves towards domestic sustainability likewise refused to be fixed entities. A process of 'identity making' formed a crucial part of participants' engagements with the GreenHomes program, with significant consequences for outcomes. For example the provision of recycling bins by participants' local government authority, rather than being a static piece of moulded plastic that allowed individual environmental values to be expressed (for example 'I care therefore I recycle'), introduced a new entity into domestic landscapes: an entity with specific needs and demands that had to be met. Some participants expressed nascent ethical obligations towards it (that is, 'must be filled higher than the other "landfill" bin'; 'put out and empty at the right time'; and 'fill it with the "right stuff"'). Along similar lines, shower timers given away to participants of GreenHomes were key to participants' notions and practices of 'water discipline', which resonated with and into other water use areas such as washing clothes and running the tap when brushing teeth. This is because, rather than being mute objects that act to express pre-cognised will, such objects were instrumental in a form of 'world-building' (Chapman 2004) stimulated by involvement with GreenHomes, wherein participants were able and willing to evaluate the functioning and meanings of their current domestic set up.

That new 'stuff' can change aspects of one's worldview and behaviour in some small way and then spread across different behavioural domains is neither surprising nor a new finding in itself (for example see Thogersen and Olander 2003). Neither is the claim that individuals are responding through affective and embodied means to objects such as recycling and compost bins or shower timers.

Yet there is another side to this story of the creation of a new 'techno-ethical' praxis, as I have called it elsewhere (Hobson 2006c). The flipside is that, while fostering some positive practices and ethics, the materialities brought to the fore by programs such as GreenHomes also appeared to circumvent, for some participants at least, further engagements with questions of sustainable consumption and sustainability in general.

That is, some participants commented that these new objects functioned as symbols of their own apparent 'greening', and reminded them on a daily basis that they were already doing enough to help the environment. Here, energy saving light bulbs, for example, appeared to give participants an (arguably exaggerated) sense of their own contribution to resource saving, allowing them to rest easy that they were doing 'all I can'. For others an acute awareness of an alleged injustice at the heart of the domestic sustainable consumption agenda was embodied in certain objects: a finding also echoed in my UK research into the Global Action Plan's Action at Home program in the late 1990s (for example see Hobson 2002, 2003). For example a feeling of 'why should I change my behaviour when government isn't taking the lead and no one else is doing anything' was cited in both the UK and Australian research as a reason for rejecting certain new materialities deemed ugly or sub-standard in terms of function (for example compact fluorescent light globes); too inconvenient, time-intensive or too much of a 'sacrifice to my lifestyle' (for example car pooling) or the responsibility of other people (for example heating systems: see also Sauter and Watson 2007).

Both sets of reasons for participants limiting their behaviour changes undoubtedly represent different issues and responses. One is about a distorted perception of the impacts of one's own practices; the other a form of active resistance. Taken together, however, they signal a paradox within and raise questions about the emphasis placed by current approaches to domestic sustainable consumption on the getting and using of green goods as a valid pathway to sustainability. That is, all of the above were no doubt the outcome of the many difficulties of trying to make behavioural and material changes in the (often) shared, complex and highly personalised space of the home. But they also suggest that the material goods so crucial to the GreenHomes and Action at Home programs are not simply the means to enact (or not) one's own or others' good intentions. Rather these objects and the practices they in part create and sustain were essential in forging a nascent and highly specific form of 'environmental citizen': one that engages with prevailing discourses and governmental interventions of domestic environmental responsibility as personal, pragmatic, small-scale but ultimately adequate in terms of cumulative impacts. That is, a form of praxis that realises in part Slocum's (2004) 'responsible and calculating citizen', but not in an ideal or unaltered form due to the presence of the slippage and resistance to change detailed above.

Indeed the prevailing discourse and materialities of 'doing your bit' at home do appear to draw a loose but identifiable boundary around the practices deemed acceptable and worthwhile in terms of improving environmental outcomes. The majority of participants in Australia and the United Kingdom expressed the opinion that taking part in sustainable consumption programs like Action at Home was definitely preferable to overt environmental political action, which was seen as controversial, oppositional and ultimately unconstructive. Research participants indeed reported perceiving a clear distinction between 'those greenies out there', protesting, 'chaining themselves to trees' and causing disruption that benefits nobody. Obviously these were stereotypes, and particular individuals held up as exemplary of the 'greenies out there' differed between countries, given their unique histories of environmental movements and protests (for example see Hutton and Connors 1999). Deeming their practices of, for example, recycling or fitting a water-saving device in the toilet, different, participants saw themselves as more sober, law-abiding and constructive in terms of achieving positive environmental outcomes.

Thus as recent writings about green governmentality suggest, seemingly pragmatic interventions to foster responsible environmental citizens in the home are essentially evacuated of and indeed in some cases in opposition to a particular form of environmental politics. Here, environmental culpabilities are worked up and realised through the stimulus of more material goods, thus rehearsing the logic of the capitalist political economy (Maycroft 2004) that undergirds the manifold interventions which governmentality scholars identify. In short, the modern home complete with 'green stuff' is not a space where one resists the unsustainabilty of modernity at its roots. Rather it becomes a space to enact an environmental citizenship that focuses on the discourse of 'doing your bit'.

Where, then, does this leave the question of how to forge effective domestic sustainable consumption practices? As suggested above, an emphasis on household ecological modernisation does little to address the Rebound Effect and the seeming limits to dematerialising current incarnations of environmental citizenship. But hoping to enrol individuals into broader environmental-political actions also has a seemingly restricted appeal. Indeed even if – as I have argued elsewhere (Hobson 2003) – the overt resistance to making lifestyles changes outlined above can be considered a form of political praxis, it does little in the way of encouraging widespread behaviour change in current socio-political climes.

Another potential pathway is argued by some to exist at the intersection of environmental politics and sustainable materialities. That is, overlapping domestic sustainable consumption interventions with a 'strong democracy' agenda by collectivising technologies and spaces of behaviour change, a move that could in theory overcome some of the feelings of acting in isolation mentioned by my research participants, and could potentially create collective outcomes greater than the sum of individualised actions. Taking up this line of argument, Steven Hoffman and Angela High-Pippert (2005: 399) suggest that programs such as community energy projects 'might well perform the sort of civic function ... a means for political activity on the part of the broad mass of citizens who join not just for social interaction but also to be actively involved in the making of public policy'.

While not refuting these claims per se, I want to explore further one assumption underpinning this agenda of community mobilisation, which has gained seemingly unstoppable momentum around a broad range of social and policy issues (see Dahlstedt 2008, see also Defilippis, Fisher and Shragge 2006 on this broader governance agenda). That is, the notion that the trajectory of individualising environmental practices by framing them as predominantly an issue of personal and home-based consumption must, first and foremost, be overcome if any form of sustainability is ever to be achieved. For one, there are the well-argued shortcomings with the ideals and possibilities that jumping scale to the level of 'community' represents *the* site of meaningful social action (Herbert 2005). However the issue I want to take up here is the presupposition that the form of environmental citizenship outlined in the previous section represents a manifestation of social dysfunction that must be surpassed.

Such a stance gives inadequate credence to how individualising supposed environmental responsibilities is not just a distortion of the polis that can be fixed but is symptomatic of governance in modernity, a point well made by Foucault and those employing his work on governmentality. Thus rather than assuming that sustainability will simply ensue once we have collectivised our environmental intentions and practices, the question could be asked: are there ways in which the picture presented above from my past research might be reconsidered as creating space for more ongoing and positive domestic sustainable consumption praxis? In short, is all lost for domestic sustainable consumption behaviour change in its current form?

On a Personal Environmental Politics and the 'Creative Grammar' of Domestic Sustainability

As outlined above and as reported in other research, there is now a general tendency of individuals in countries such as the United Kingdom to demarcate the home as the main space where pro-environmental action takes place (see Macnaghten 2003). Indeed in terms of public attitudes, while environmental problems are perceived to be getting worse and a cause for concern (for example The Climate Institute 2008), there is less uptake of actions like signing petitions or taking part in protests (see Wall 1995). This trend is often framed as problematic by, for example, non-government organisations and other 'activist-focused' groups. Perceiving environmental politics in terms of achieving the normative goals of collective action and dedication by all to this particular cause (Macnaghten, 2003), they see any retreat from such a position as detrimental to collective environmental outcomes.

However Phil Macnaghten (2003) contends that such framings of what constitutes a sound environmental praxis rarely articulate with the broader socio-economic and political trends in which environmental concerns are embedded. As he observes: 'the question as to how environmental concerns are tied up with the emergence of this apparently more individualized and globalised society has received little attention' (Macnaghten 2003: 68).

As Macnaghten contends – and this point is certainly supported by my own empirical research and that of others (for example Bickerstaff and Walker 2001) – individuals are reportedly less concerned with abstract issues (climate change, diminishing biodiversity) and more with environmental problems that directly impact on themselves and their families, such food scares and pollution. These concerns are played out in the context of both a reported lack of trust in the 'expert' institutions offering advice on, or supposedly protecting us all from exposure to, environmental harm (see Wynne 1996), and diminished interactions with formal political actors and spaces, resulting in 'restricted political engagement based on personal identity often expressed in lifestyle bodily practices' (Macnaghten 2003: 69).

In response to this alleged set of social and cultural conditions, researchers have argued again the presupposition that a particular form of politics must ensue for society to move beyond these apparent barriers to collective action. Instead, there is a need to explore how the 'political' and 'participation' might be constituted by and articulated with everyday embodied practices. That is, we might challenge the ways in which 'the decline in collective participation and a rise in individualized or micro-political participation' (Li and Marsh 2008: 249) is too often segued with a 'liquid, modern consumerist version of the art of life' evacuated of the 'common good' ('think global, act local' and so on) and centred around the individual's 'successful life' (Bauman 2008: 77). In short, we need to rethink what a valid environmental politics might look like under the conditions of modernity (for example Szerszynski 2007).

So What Does 'Environmental Politics' Entail?

In scholarly discourse politics as a manifestation of power is often conceptualised as a set of practices where pre-defined interests, represented by a set cast of actors, tussle over ideologies and resources through institutionalised and also ad hoc means (Allen 2003). With the term originating from the Greek word 'polis' (city state), it intrinsically denotes a unit of action and concern that exceeds the individual and the quotidian. However other definitions abound as political philosophers and geographers, among many, attempt to define a broader and more inclusive sense of the political (for example Mouffe 2005, Amin and Thrift 2007). While there is not the space here to revisit these arguments in full, one contention offers a potentially insightful view of politics in terms of articulating with prevailing individualised and consumer-focused discourses and practices of environmental citizenship, as detailed above. For example Dikeç (2005: 172) contends that:

> space becomes political in that it becomes the polemical place where a wrong can be addressed and equality can be demonstrated ... The political, in this account, is signalled by this encounter as a moment of interruption, and not by the mere presence of power relations and competing interests.

Arguably such 'moments of interruption' are present in the experiences of GreenHomes and Action at Home participants. That is, research participants reported having 'ah ha' moments when they made connections between practices they had previously not given much thought to and their wider environmental and socio-political outcomes, such as running taps while brushing teeth and throwing away still edible food (see Hobson 2003 for further details of this argument). As a result these participants reported feeling impelled to change these behaviours, as much out of an ethical sense as a feeling of 'I can do that, it's not so hard', thus arguably speaking of the 'moment of interruption' Dikeç speaks of.

As such, the political here is not confined to one particular space or praxis. Indeed some thinkers have advocated the promotion and cultivation of personal behaviour as thoroughly political, in the sense of being 'the polemical place where a wrong can be addressed' (Dikeç 2005: 172). For example, arguing in the political context of his time, the early twentieth-century Chinese thinker Zhang Shizhao advocated that personal behaviour does not transcend politics (see Jenco 2008). Thus thinking that effective action can only be taken with others leads to the incorrect assumption that our individual contributions do not much matter. Along similar lines, some of Foucault's later writings focused on how one could develop one's own ethics as a way of resisting the governmental imperatives outlined so thoroughly in his earlier work. In this writing he worked to reintroduce the ethos of Hellenistic practices into repertoires of political action and resistance. Fundamentally he argued that individuals could form ways of being or 'practices of freedom' that could contend with contemporary techniques of power that aim to discipline and suppress (Myers 2008). How this might be done, according to Foucault, involved various forms of self-reflection and self-directed practices that constantly ask 'What do I aspire to be?' (see Cordner 2008).

Critics of such positions argue that dangers lie therein. For example the presumed self-absorption of such a form of politics can lead to forgetting the needs of others and the context of operation, creating a plethora of 'kingly' individuals who have little thought for the needs of others. Or as Leigh Jenco (2008: 218) has put it: 'Just as individuals cannot create their own personal languages, neither can they initiate unilateral political change without taking into account the existing "grammar" of the community that gives meaning to their action and words.'

Indeed, from my own research, it is clear that relying on individuals' 'moments of interruption' (Jenco 2008: 218) as a form of meaningful politics is not without problems, for the reasons to which Jenco alluded; and also because these moments soon sink back down into habit and 'practical consciousness' (see Hobson 2003), providing little further impetus to keep asking 'What do I aspire to be?', following on from Foucault.

However Jenco's idea of 'grammar' merits further consideration. Using this term, Jenco is no doubt referring to the shared rules of communication and understanding in which we all partake, without always being consciously aware of doing so. Indeed without some adherence to these norms we would not be able to comprehend what others are talking about or make ourselves understood.

Translating this into political practice, this suggests that the range of available personal-political actions is circumscribed by discourses of accepted and acknowledged praxis, outside of which their impacts and meanings are lost – and indeed outside of which the ability to be labeled as political becomes problematic. Such a proposition suggests that a meaningful environmental politics can do little except trade in the existing grammars of received praxis, politics and participation. However, as with language itself, the rules and norms of grammar are open to variation, evolution and resistance. Thus as with the room for resistance and change that governmentality scholars have recently discussed (see above) there is room within existing 'grammars of praxis' for variation and 'slippage'.

What then has this to do with domestic sustainable consumption and debates over the appropriate site for and forms of environmental citizenship that must be fostered to create less environmentally bleak futures? My suggestion here is that there may be scope for harnessing such slippage, mobilising existing grammars in ways that move towards a constructive and positive slippage, such as fostering a creative grammar of household sustainable consumption that constitutes it as a thoroughly political act; that is, where there is a moment, or ongoing moments, of interruption that can take the needs and presence of other people and non-human entities into account and can, in turn, be acknowledged by others.

In terms of the materialities discussed in this chapter, such praxis is not always possible because, as outlined above, much of the 'stuff' of household greening was reported to close down further 'interruptions', that is, once the light bulbs had been fitted or the shower timer use become just another daily habit. How then to foster creative materialities that allow ongoing engagements with questions of sustainability and a 'successful life'? As I have discussed elsewhere (Hobson 2008), redesigning living spaces to reflect a different form of success (to deploy Bauman's (2008) term) is one possible way to proceed. For example households in Australia have worked with community facilitators to raise the question of what they value in the everyday. Rather than this being the acquisition of more objects or meeting pressing deadlines, it was shown that playfulness, time and communication with others were considered markers of a successful or good life. In response some people went about altering their home spaces to reflect these sentiments, such as removing their front lawn fences and putting sofas out the front to 'reclaim their street' as a place of belonging and not just a highway that cars zoom down and children no longer play in: a move that some reported as constantly reminding them of what was important to them, reiterating Foucault's question of 'Who do I want to be?'

Others have talked about the role that design can play in such moves, whereby playfulness and creativity become part of the materialities of sustainable consumption. Here design is not so much about producing the end product but about rethinking the purpose and manner of how certain values are delivered. As De Bono (2000: 140) stridently puts it: 'Most of the major problems in the world will not be solved by more analysis. There is a need for design.' This agenda is fundamentally different from that of domestic ecological modernisation, discussed above, as it is not just about making objects more efficient per se. Rather it is using the heuristic potential of materialities. Or as Cameron Tonkinwise (2003: 10) has put it, 'dispens[ing]with the technical focus on completed products. It will require that design be more receptive to incompletion, to products-in-time, to things changing, in ways that cannot be pre-empted'.

Cameron Tonkinwise's suggestion has resonance with the aim of this chapter. That is, to think about a form of environmental politics and participation that is neither slavishly adherent to the 'responsible and calculating' consumer-citizen discourse, nor the activist-protestor that Macnaghten (2003) critiques. Rather there may be potential for a form of personal and domestic environmental politics wherein green materialities re-invoke 'moments of interruption' to constitute an evolving creative grammar of praxis.

Concluding Remarks

As this chapter has pointed out, recent research – and a re-visitation of my own work through the lens of green governmentality – suggests there are rationalities and technologies at work that both activate and circumscribe the form and scope of contemporary domestic sustainable consumption practices. Yet such circumscription does not automatically presuppose an apolitical subject, or that 'environmental citizens' are uncritically moulded into the embodiment of their ideal form. Rather it has been proposed that within the seemingly self-referential and 'stuff-focused' trajectory of contemporary sustainable lifestyles, the potential for a meaningful material environmental politics just might exist. To some it may seem evasive (and a bit of a cop out) to not sketch out more concretely what this politics might look like. But, as stated above, such conjecture was not the stated aim of this chapter, nor is it appropriate, given the methodologies implicated. What is clear however is that social researchers have a key role to play in the development of such a materialised creative grammar. There is potential to carry out action research, along with designers and households, to explore the possibilities of populating the home space with the 'incomplete products' which Tonkinwise discusses, which could constitute an environmental politics that articulates with existing grammars of action in creative ways. What form and purpose they may take is hard to predict out of context: ideas such as 'extended producer responsibility' or systems of resource-use feedback are just some of the possibilities already in play, and indeed the nature of the shared creativity required for such endeavours makes pre-emption prescriptive rather than helpful. But certainly a fundamental 'take home' point from this chapter is the key role researchers can play in not shutting down the debate on household sustainable consumption before it has got really interesting: that is, not assuming that the everyday practices associated with materialities of domestic sustainable consumption are inherently devoid of a worthwhile form of personal and material environmental politics.

References

Allen, J. 2003. *Lost Geographies of Power*. Oxford: Blackwell.

Amin, A. and Thrift, N. 2007. On being political. *Transactions of the Institute of British Geographers*, 32(1), 112–115.

Anderson, B. and Tolia-Kelly, D. 2004. Matter(s) in social and cultural geography. *Geoforum*, 35(6), 669–674.

Bauman, Z. 2008. *The Art of Life*. Cambridge: Polity Press.

BBC. 2006. Australia celebrates baby boom. *BBC News* [Online, 2 June]. Available at: http://news.bbc.co.uk/2/hi/asia-pacific/5040582.stm [accessed: 24 September 2010].

Bickerstaff, K. and Walker, G. 2001. Public understandings of air pollution: the 'localisation' of environmental risk. *Global Enviromental Change*, 11(2), 133–145.

Burgess, J., Bedford, T., Hobson, K. and Harrison, C. 2003. (Un)sustainable consumption, in *Negotiating Environmental Change: New Perspectives from Social Science*, edited by F. Berkhout, M. Leach and I. Scoones. Cheltenham: Edward Elgar, 261–291.

Chapman, A. 2004. Technology as world building. *Ethics, Place and Environment*, 7(1–2), 59–72.

Cordner, C. 2008. Foucault, ethical self-concern and the other. *Philosophia*, 36(4), 593–609.

Cupples, J. and Ridley, E. 2008. Towards a heterogeneous environmental responsibility: sustainability and cycling fundamentalism. *Area*, 40(2), 254–264.

Dahlstedt, M. 2008. The politics of activation: technologies of mobilizing 'multiethnic suburbs' in Sweden. *Alternatives*, 33(4), 481–504.

De Bono, E. (2000). *New Thinking for the New Millennium*. London: Penguin Books.

Defilippis, J., Fisher, R. and Shragge, E. 2006. Neither romance nor regulation; re-evaluating community. *International Journal of Urban and Regional Research*, 30(3), 673–689.

Department of Climate Change and Energy Efficiency (2010) Individuals. Australian Government Department of Climate Change and Energy Efficiency http://www.climatechange.gov.au/what-you-can-do/individual.aspx. Accessed 29 September 2010.

Dikeç, M. 2005. Space, politics, and the political. *Environment and Planning D*, 23(2), 171–188.

Foucault, M. 2004. *Society Must Be Defended*. London: Penguin.

Geller, E.S. 1992. It takes more than information to save energy. *American Psychologist*, 47(10), 814–815.

Gittins, R. 2009. Emission impossible: the sad truth. *Sydney Morning Herald* [Online, 24 February]. Available at: http://business.smh.com.au/business/

emission-impossible-the-sad-truth-20090224-8gsv.html [accessed: 25 February 2009].

Herbert, S. 2005. The trapdoor of community. *Annals of the Association of American Geographers*, 95(4), 850–865.

Hobson, K. 2001. Sustainable lifestyles: rethinking barriers and behaviour change, in *Exploring Sustainable Consumption: Environmental Policy and the Social Sciences*, edited by M.J. Cohen and J. Murphy. Oxford: Pergamon Press, 191–209.

Hobson, K. 2002. Competing discourses of sustainable consumption: does the 'rationalisation of lifestlyes' make sense? *Environmental Politics*, 11(2), 95–120.

Hobson, K. 2003. Thinking habits into action: the role of knowledge and process in questioning household consumption practices. *Local Environment*, 8(1), 95–112.

Hobson, K. 2004. Sustainable consumption in the United Kingdom: the 'responsible' consumer and government at 'arm's length'. *Journal of Environment and Development*, 13(2), 121–139.

Hobson, K. 2006a. Environmental psychology and the geographies of ethical and sustainable consumption: aligning, triangulating, challenging? *Area*, 38(3), 292–300.

Hobson, K. 2006b. Environmental responsibility and the possibilities of pragmatist-orientated research. *Social and Cultural Geography*, 7(2), 283–298.

Hobson, K. 2006c. Bins, bulbs and shower timers: on the 'techno-ethics' of sustainable living. *Ethics, Place and Environment*, 9(3), 335–354.

Hobson, K. 2008. Reasons to be cheerful: thinking sustainably in a (climate) changing world. *Geography Compass*, 2(1), 199–214.

Hoffman, S.M. and High-Pippert, A. 2005. Community energy: a social architecture for an alternative energy future. *Bulletin of Science, Technology and Society*, 25(5), 387–401.

Hutton, D. and Connors, L. 1999. *A History of the Australian Environment Movement*. New York: Cambridge University Press.

Huxley, M. 2008. Space and government: governmentality and geography. *Geography Compass*, 2(5), 1635–1658.

Jelsma, J. 2003. Innovating for sustainability: involving users, politics and technology. *Innovation*, 16(2), 103–116.

Jenco, L.K. 2008. Theorists and actors: Zhang Shizhao on 'self-awareness' as political action. *Political Theory*, 36(2), 213–238.

Kaika, M. 2004. Interrogating the geographies of the familiar: domesticating nature and constructing the autonomy of the modern home. *International Journal of Urban and Regional Research*, 28(2), 265–286.

Li, Y. and Marsh, D. 2008. New forms of political participation: searching for Expert Citizens and Everyday Makers. *British Journal of Political Science*, 38(2), 247–272.

Luque, E. 2005. Researching environnmental citizenship and its publics. *Environmental Politics*, 14(2), 211–225.

McKee, K. (2008). Transforming Scotland's public sector housing through community ownership: the reterritorialisation of housing governance. *Space and Polity*, 12(2), 183–196.

Macnaghten, P. 2003. Embodying the environment in everyday life practices. *The Sociological Review*, 51(1): 63–84.

Maniates, M. 2002. Individualization: plant a tree, buy a bike, save the world?, in *Confronting Consumption*, edited by T. Princen, M.F. Maniates and K. Conca. Cambridge, MA: MIT Press, 343–366.

Maycroft, N. 2004. The objectness of everyday life: disburdenment or engagement? *Geoforum*, 35(6), 713–725.

Merriman, P. 2005. Materiality, subjectification, and government: the geographies of Britain's Motorway Code. *Environment and Planning D*, 23(2), 235–250.

Mouffe, C. 2005. *On The Political*. New York and Abingdon, UK: Routledge.

Myers, E. 2008. Resisting Foucauldian ethics: associative politics and the limits of the care of the self. *Contemporary Political Theory*, 7(2), 125–146.

Nelson, L. 1999. Bodies (and spaces) do matter: the limits of performativity. *Gender, Place and Culture*, 6(4), 331–353.

Putnam, R. 2000. *Bowling Alone: The Collapse and Revival of American Community*. New York: Simon and Schuster.

Raco, M. 2003. Governmentality, subject-building, and the discourses and practiecs of devolution in the UK. *Transactions of the Institute of British Geographers*, 28(1), 75–95.

Rutherford, S. 2007. Green governmenality: insights and opportunities in the study of nature's rule. *Progress in Human Geography*, 31(3), 291–307.

Rutland, T. and Aylett, A. 2008. The work of policy: actor networks, governmentality, and local action on climate change in Portland, Oregon. *Environment and Planning D*, 26(4), 627–646.

Sauter, R. and Watson, J. 2007. Strategies for the deployment of micro-generation: implications for social acceptance. *Energy Policy*, 35(5), 2770–2779.

Science and Technology Committee and The House of Lords UK. 2005. *Energy Efficiency. Volume 1: Report*. London: The Stationary Office Limited.

Shove, E. 2003. *Comfort, Cleanliness and Convenience: The Social Organization of Normality*. Oxford: Berg.

Slocum, R. 2004. Consumer citizens and Cities for Climate Protection campaign. *Environment and Planning A*, 36(5), 763–782.

Staats, H.J. and Harland, P. 1995. The EcoTeam Program in the Netherlands. *Study 4 : A longitudinal study on the effects of the EcoTeam Program on environmental behavior and its psychological backgrounds*. Leiden: Centre for Energy and Environmental Research, Leiden University.

Staeheli, L.A. and Mitchell, D. 2004. Spaces of public and private: locating politics, in *Spaces of Democracy: Geographical Perspectives on Citizenship,*

Participation and Representation, edited by C. Barnett and M. Low. London: Sage Publications, 147–160.

Szerszynski, B. 2007. The post-ecologist condition. *Environmental Politics*, 16(2), 337–355.

The Climate Institute 2008. *Climate of the Nation: Australian Attitudes to Climate Change and its Solutions*. Sydney, New South Wales: The Climate Institute.

Thogersen, J. and Olander, F. 2003. Spillover of environment-friendly consumer behaviour. *Journal of Environmental Psychology*, 23(3), 225–236.

Tonkinwise, C. 2003. *Interminable Design: Techné and Time in the Design of Sustainable Service Systems*. Proceedings of 5th European Academy of Design Conference — Techné, Design, Wisdom. Barcelona. [Online]. Available at: www.ub.es/5ead/PDF/8/Tonkinwise.pdf [accessed: 24 September 2010].

Troy, P. (ed.) 2008. *Troubled Waters: Confronting the Water Crisis in Australian Cities*. Canberra: ANU E-Press.

Verplanken, B. and R. W. Holland, R.W. 2002. Motivated decision making: effects of activation and self-centrality of values on choices and behavior. *Journal of Personality and Social Psychology*, 82(3), 434–447.

Walker, G. and Cass, N. 2007. Carbon reduction, 'the public' and renewable energy: engaging with socio-technical configurations. *Area*, 39(4), 458–469.

Wall, G. 1995. Barriers to individual environmental action: the influence of attitudes and social experiences. *The Canadian Review of Sociology*, 32(4), 465–489.

Whitmarsh, L. 2009. Behavioural responses to climate change: asymmetry of intentions and impacts. *Journal of Environmental Psychology*, 29(1), 13–23.

Wood, G. and M. Newborough, M. 2003. Dynamic energy-consumption indicators for domestic appliances: environment, behaviour and design. *Energy and Buildings*, 35(8), 821–841.

Worldwatch Institute. 2008. *The State of Consumption Today*. [Online]. Available at: www.worldwatch.org/node/810#1 [accessed: 9 Febraruy 2010].

Wynne, B. 1996. May the sheep safely graze? A reflexive view of the expert–lay knowledge divide, in *Risk Environment and Modernity: Towards a New Ecology*, edited by S. Lash, B. Szerszynski and B. Wynne. London: Sage Publications, 44–83.

Discussion: Governance and Citizenship at Home

Aidan Davison

The chapters in this section focus on the practices and aspirations of householders in relation to governance for sustainability. Although diverse and innovative in their contemporary lines of inquiry, these chapters are united by an effort to make sense of a venerable question: namely, where does 'social structure' end and 'individual agency' begin in the conduct of politics? Admittedly these terms are not themselves prominent here, attesting among other things to the waning of neo-Marxian analysis in human geography. And there is running through much material presented an undercurrent of post-structuralist interest in the constitution of power and of the heterogeneity of agency. The themes of governance and citizenship developed nonetheless tackle the familiar ontological problem of how best to conceive of the relationship between individual lives and social worlds. Laudably the authors recognise that this is a problem demanding both rich empirical detail and inclusive normative argument. In what follows, then, I reflect upon the productive encounter of the four contributions to this section with an eye to both new opportunities and perennial concerns.

The authors broadly share a critique of the one-dimensional rationality found in many governmental approaches to sustainability; a critique most fully articulated by Kersty Hobson under the heading here of 'green governmentality' (see also Hobson 2002). While not denying the positive effects of governance aimed at aligning environmental objectives with rational consumer choices through education campaigns, price signals and regulatory constraints, this critique asserts the 'multiple rationalities' that Watson and Lane identify at play in consumption practices. As is argued in this collection more generally, conventional binaries between public and private, citizen and consumer obscure the breadth and depth of concerns at work in acts of consumption, especially those acts wrapped up in the practices of homemaking. Neat governmental agendas for securing sustainability are apt to become unpicked and tangled up in the complexity of household consumption practices. As a result governmental strategies often steer clear of unruly consumer behaviours, as can be seen in Matt Watson and Ruth Lane's study of the way in which formal waste management strategies overlook flows of second-hand goods. These flows are a familiar feature of everyday life in advanced consumer societies – the waste of the wealthy not surprisingly constituting rich pickings – yet confound bureaucratic accounts of economic exchange, managerial notions of waste and neo-liberal narratives about consumer motives.

Although contributors to this section critique the instrumentalism of green governmentallty and point to a disjuncture between governmental programs and household practices, they are aware that governance practices are no less multiple and internally conflicted than domestic practices. In their introduction to this volume, Ruth Lane and Andrew Gorman-Murray observe that 'neo-liberal governance approaches stress the roles and responsibilities of individual decision makers'. Andy Scerri fleshes out this argument by tracing neo-liberal governmental reliance on the discourses and practices of the 'stakeholder citizen'. Efforts to govern stakeholder citizens emphasise the capacity of self-governing individuals to achieve common goods, such as sustainability, through autonomous consumer choices and 'lifestyle' preferences. At the same time, such governance downplays the role of stakeholder citizens in shaping social institutions, representing goals such as economic growth and technological globalisation as apolitical imperatives and advocating what Scerri calls 'institutional voluntarism'. Indeed, in keeping with the analogy between stakeholder citizens and the shareholders of capitalist corporations, representations of self-regulating 'citizen consumers' are aligned with representations of self-regulating 'corporate citizens' in neo-liberal discourses of sustainable development.

The 'rebound effect' noted by Hobson, by which increased resource use efficiencies are offset or outstripped by overall increases in consumption, is presented within governance frameworks as an unintended side-effect of rational institutional agendas with origins in the vagaries of consumer behaviour (for example Hertwich 2005). Andy Scerri's analysis, however, implies that this effect is better understood as a predictable consequence of the coupling of sustainability initiatives and corporate capitalism in the logic of neo-liberal governance. Matt Watson and Ruth Lane make a similar point in suggesting that questions about sustainable governance arrangements for waste management need to be understood in the context of the potential for these arrangements to conflict with 'dominant neo-liberal ideologies of public policy' by promoting reduced consumption. For example while householders may see good reasons for conferring the gift of second life on things – or for making do with last year's model or for sharing with neighbours – the question of whether these reasons match up with institutional constructions of the common good is often left to deliberations about efficient management and related attempts to align the interests of consumers and citizens.

The observation that green governmentality encourages stakeholder citizens to accept personal responsibility for environmental impacts while discouraging them from engaging with issues of collective responsibility sheds light on the efforts of governments in Australia and other advanced consumer societies to promote sustainable domestic consumption. Kersty Hobson's argument that domestic consumption of 'green goods', such as shower timers and energy efficient lighting, is not just the outcome of such governance but also constitutive of it, is insightful. Her research indicates that the material embodiment of claims about sustainability in the home through green goods is bound up with governmental practices and discourses of the empowered 'environmental citizen' who discharges their responsibility to the planet and to the future via consumer preferences.

Drawing on a survey of householders in Melbourne, Scerri concludes that the conflicted nature of neo-liberal efforts to govern in the name of sustainability is becoming increasingly untenable as citizen-consumers translate their environmental concerns into 'green' value commitments that challenge the basic tenets of neo-liberalism's version of the good life. The implication here is that householders are increasingly likely to seek opportunities to express their normative vision of sustainability by challenging prevailing economic and technological norms. In effect, then, Louise Crabtree picks up where Scerri leaves off in her consideration of the potential for up-scaling of forms of householder activism that pursue alternative practices of home ownership. Taking cohousing and Community Land Trusts as her focus, Crabtree documents a movement from small resident-driven initiatives to larger projects driven by governmental and non-governmental organisations. She argues that in the process of up-scaling, active community participation in these projects is at risk of being supplanted by private economic interests, specialist expertise and governmental agendas. Given the often slow, fraught and complex nature of participatory decision making, Crabtree notes also the risk that technological design strategies will be regarded as an efficient tool for redirecting lifestyles toward sustainability outcomes without need for conscious deliberation and debate, and thus may stand in for citizen involvement in larger-scale cohousing and Community Land Trust projects.

Matt Watson and Ruth Lane similarly address the issue of up-scaling of household sustainability initiatives that presently sit outside of green governance frameworks, so as to bring about changes to these frameworks. While they see potential for some articulation between circuits of reuse and waste management regimes, Watson and Lane point out two significant barriers to up-scaling, in addition to the broader concern canvassed above about neo-liberal growth agendas. First, the present restricted framing of waste management governance as an environmental agenda fits poorly with the broad range of social and environmental motives that give impetus to household practices of reuse. Second, the present engagement of waste management strategies with household practices is mostly limited to well-established material networks linking homes to landfill sites. These material relations need to be diversified considerably if they are to encompass the 'bewildering array' of actors, places and things bound together in diverse networks of reuse. I take it from this that Watson and Lane are, like Crabtree, concerned that up-scaling of household sustainability initiatives, unaccompanied by overt efforts at institutional reform, will likely see households become even more vulnerable to regimes of neo-liberal discipline in the form of the 'environmental stakeholder citizen'.

Discussion about whether citizen-based sustainability initiatives can and ought to be scaled up returns us to the discourses and practices of neo-liberal citizenship. In addition to the governmental reliance on self-governing stakeholders observed by Scerri, neo-liberal reform is also linked to what Luke Desforges, Rhys Jones and Mike Woods (2005: 440) describe as:

a rescaling downwards of the performance of citizenship to more local contexts as part of the transition to a new mode of governmentality. Whereas the mode of 'managed liberalism' that was dominant in most post-war advanced liberal democracies prioritised the 'national citizen' in its emphasis on the security of social, political and economic rights at the national scale, the new mode of 'governing through communities' shifts the emphasis to the practice of responsibilities by 'active citizens' in sub-national communities.

Localisation of the practices of citizenship, and resulting divergence of questions of citizenship from questions to do with national and transnational institutions of neo-liberal governance, is evident in the fate of the United Nations' 1992 action plan for sustainable development, *Agenda 21*. Rightly lambasted for failing to bring about meaningful action from the national governments that produced it, *Agenda 21* is notable also for the influence it has had at the level of municipal government and for its role in launching the twin concepts of sustainable consumption and sustainable production. Not coincidentally, these concepts have been invaluable to nation-states in their attempt to constitute governance for sustainability through the voluntary action of consumers and corporations.

Rather than suggesting that the only political response to the localising and individualising of citizen action in the name of sustainability is up-scaling, Hobson points to an intriguing, if as yet indistinct, alternative political praxis of 'interruption'. Noting that 'jumping scale to the level of "community" as a site for action is not without considerable problems', Hobson asserts that far from being a stable or completed form of environmental citizenship, the responsible homeowner constituted by green governmentality is a necessarily fragile and partial achievement. Resisting the conclusion that 'effective action can only be taken with others', Hobson points to under-regarded possibilities within the material field of domestic consumption for interrupting the performance of the environmental stakeholder citizen. Complementing Crabtree's concern about the use of green design as a tool of technocratic control, Hobson picks up on radical discussion within design disciplines to imagine a domestic realm shaped by a technological ethos that values contingency, uncertainty and lightness of touch.

In concluding, I follow Hobson's lead in gesturing towards a strategy for deepening the political significance of household sustainability initiatives that does not rely on linear notions of scaling up; a strategy that seems to me could be integrated with Hobson's interest in the design 'grammar' of domestic action (see for example Latour 2002). This strategy lies in presently burgeoning experiments in human geography and cognate fields that aim at 'a different relational and topological way of conceptualising space' (Desforges, Jones and Woods 2005: 443). This interest in 'relational space' invites inquiry to overcome what Hobson rightly observes is the disarticulation of global abstractions about sustainability and the intimate, embodied preoccupations of domestic life. I agree with Martin Jones's (2009: 494) recent caution about the need for a 'moderate relationalism' which asserts that 'mobility and fluidity should not be seen as standing in opposition to territories and we should, therefore, *not* be forced to adopt a "networks versus territories" scenario'. There is no doubt in my mind that modernity is a paradoxical affair. Neo-liberal governance both undermines and relies upon territorial social formations, including the territory of the neo-liberal home so neglected, until recently, by researchers. With the caveat that contemporary social relations are liquid and solid, unprecedented and historical, then, I suggest that topological accounts of how householders are drawn together and apart in multiple networks of proximity operating at multiple scales promise potentially transformative ways of documenting the nexus of governance and citizenship in practices of household sustainability.

References

Desforges, L., Jones, R. and Woods, M. 2005. New geographies of citizenship. *Citizenship Studies*, 9(5), 439–451.

Hertwich, E.G. 2005. Consumption and the rebound effect: an industrial ecology perspective. *Journal of Industrial Ecology*, 9(1–2), 85–98.

Hobson, K. 2002. Competing discourses of sustainable consumption: does the 'rationalisation of lifestyles' make sense? *Environmental Politics*, 11(2), 95–120.

Jones, M. 2009. Phase space: geography, relational thinking, and beyond. *Progress in Human Geography*, 33(4), 487–506.

Latour, B. 2002. Morality and technology: the end of the means. Translated by C. Venn. *Theory, Culture & Society*, 19(5), 247–260.

Chapter 12

Conclusion: Tackling the 'Missing Scale' in Environmental Policy

Ruth Lane and Andrew Gorman-Murray

The foregoing chapters have demonstrated that household sustainability is a complex issue that requires thoughtful discussion from multiple perspectives. Indeed throughout this collection we have encouraged a dialogical approach. On the one hand our aim has been to bring researchers from human geography and cultural studies into a productive dialogue around the material geographies of household sustainability. There has been enthusiastic cross-fertilisation of ideas and approaches shown through the chapters, which critically develop the interconnections between the material, socio-technical, cultural, embodied and political dimensions which make households function, with particular attention to how they (might) function in environmentally sustainable ways. On the other hand the actual format of the book also deployed a dialogical approach, with discussants providing commentary and reflection on the chapters in each of three sections. These discussion pieces draw out the connections between the chapters and are suggestive of possibilities for developing further work. This approach to the edited collection means that tentative conclusions have already been provided throughout the book. So in this brief 'wrap up' we want to return to the big themes and suggest key areas for advancing research on household sustainability from the perspectives of material geographies. Having encouraged dialogue, we do not want to close (it) off with a 'final' conclusion, but rather clear ground for further interrogation and debate.

This wrap up is arranged in three sections. Firstly, we return to the idea of material geographies advanced throughout the book, and further contemplate the value of this approach – or rather, this set of perspectives. We reintroduce the different approaches to materiality and stress how they together reconfigure normative ideas of 'the material world', disrupting a strict distinction between object and subject, the material and the immaterial, and instead assert that the material, cultural, objective and subjective are co-constituted. Secondly, we reflect on the importance of the often neglected household scale – the geographical focus of this collection – for activating conscious and habitual environmental sustainability. The household is a key spatial frame in relational politics around environmental action and governance, connecting national strategies and regional embeddedness with individual behaviours. Thirdly, we consider directions for further research inspired by the authors in this collection and their varied perspectives on the material geographies of household sustainability.

Material Geographies Approach(es)

The introductory chapter outlined our material geographies approach for understanding household sustainability. In conventional thought, the material – the objective, real, hard and tangible – is often opposed in a framework of binary logic with the immaterial – the subjective, ideal, soft and discursive. However 'new' approaches to materiality within human geography have moved against this dualistic understanding and have attempted to reconfigure these normative notions of 'the material world'. We also seek to move beyond this binary framing. In its place we have iterated several approaches to material geographies that, in fact, urge the disruption of such constructed dualisms as those around subject/object, material/immaterial, nature/culture and mind/body. Instead these approaches contend that the material, the cultural, the natural, the objective, the subjective and the mind and body are all co-constituted and not so easily divisible into separate ontological categories of analysis. Our opening chapter introduced and discussed three particular clustered approaches to 'new' material geographies, following a recent review by Ben Anderson and John Wylie (2009). This was a handy foundation for our thinking, and we wanted to advance these three material geographies approaches by showing their applicability for understanding household sustainability in the contemporary world.

These three approaches can be identified and summarised as material cultures, hybrid geographies and embodied geographies. Material cultures identifies the cultural embeddedness, mediation and meaning of objects; hybrid geographies describes the co-constitution of nature and culture; embodied geographies identifies the intersection of emotions, corporeality and sociality in the idea of 'the self'. While we distinguished these three reasonably distinct approaches, this framework was not picked up in such a neat manner by the authors in this collection. For instance, the hybrid geographies emerging within this collection highlight the interrelationships between householders, technology and infrastructure rather than a definite nature–culture relationship. Nature, for these authors, is primarily framed in terms of the environmental resources caught up in the water, energy and goods and materials consumed in the home. It is these that are construed as the 'nature' in need of conservation in the city, which echoes Maria Kaika's (2004) argument about 'domesticated nature' entering the modern home as flows of commoditised resources. Aidan Davison's chapter offers particular insights here in tracing the re-framing of environmental politics over time away from a focus on 'wild' nature and towards debates around eco-efficient technologies. In investing in eco-efficiency, householders are also engaged in a very localised form of resource security or 'future-proofing' that mirrors the discourse of 'urban ecological security' that is increasingly used to promote sustainable development templates for world cities (Hodson and Marvin 2009).

Overall though, it is the myriad intersections between these three approaches to material geographies that stand out rather than any one of them. Perhaps this is inevitable in empirical research at the household scale. Research findings are contingent on the specific social, cultural and geographical contexts of the households explored. In a number of contexts the current imperatives around sustainable housing find little purchase as certain types of living space are considered more desirable. For example Robyn Dowling and Emma Power highlight the importance of expansive living spaces to the cultural value of 'homeyness' among families living in large houses in the outer suburbs of Sydney – in direct contravention of the current emphasis on reducing resource consumption. Willem Paling and Tim Winter highlight the different circumstances of the emerging professional classes in the rapidly modernising city of Phnom Penh that are moving housing aspirations away from the more sustainable traditional architecture that incorporated passive cooling approaches towards 'modern' styles reliant on air-conditioning.

The various chapters demonstrate the 'messiness' that John Law (2004) argued is central to materially grounded social research, where stated values and attitudes often diverge from observable practice, and generalisations are not always appropriate. This poses specific challenges for 'scaling up' research findings to inform policy approaches. Kersty Hobson, in her chapter contribution, suggests a metaphor of 'creative grammar' to help understand the interconnections between discourse and practice around household sustainability. While existing and evolving norms and conventions can be observed, there is also much variation or 'slippage' around these, and innovation within households can both interrupt conventional practice and be acknowledged by others. This volatility highlights the potential for research around household practices to inform new policy approaches, as well as to evaluate their effectiveness.

One area of innovation is in tenure and property arrangements affecting both the built environment and moveable goods. Louise Crabtree's documentation of cooperative housing models highlights some of the differences between community-driven initiatives and more formally constituted housing cooperatives in incorporating various forms of sustainable resource use in housing design, noting that community-driven initiatives are more likely to include a wider range of sustainability principles in their design. In arguing the significance of informal channels for the reuse of second-hand goods, Matt Watson and Ruth Lane contrast the multiple interpretations of property in used goods associated with the diverse rationales and motivations around community-based initiatives, compared with the more singular assumptions about property implicit in government waste management initiatives. Related to these issues of property is the slippage between the categorisation of particular environmental resources as either products or services, as in Gay Hawkins and Kane Race's analysis of bottled water versus tap water in Bangkok. These categorisations draw attention to the role of the material form, in this case water bottles, in defining environmental resources as marketable products, rather than regulated services, and in doing so highlight the limitations of market models relying on 'citizen-consumers' for promoting real efficiencies in resource use.

The Politics of the Household Scale

In this collection we have introduced and advanced a material geographies approach with a specific purpose: to provide new insights into household sustainability. In terms of environmental policy, action and governance, as well as scholarly debates about environmental decision making, what is critical here is the spatial focus of our material analysis on the household. As Louise Reid, Philip Sutton and Colin Hunter (2010) have recently and cogently argued, the household is a 'crucible' for pro-environmental behaviours that has not been adequately conceptualised and analysed in sustainability research across a range of disciplines, including human geography. Governmental policies about sustainability are increasingly focused on the household, for instance through programs for the collection and recycling of household waste, and through price mechanisms to encourage reduced household energy and water usage. But this approach treats the household as if it was an individual unit of behavioural decision making rather than a collective of separate individuals with their own environmental behaviours and their own connections both within and beyond the household. Instead Reid and colleagues (2010) characterise the household as a unit within the *meso* scale, and contend that it is an important relational site that articulates between the *macro* scale (policy change and societal attitudes) and the *micro* scale (individual behaviour). Just as material geographies approaches disrupt binary understandings of the material and the immaterial, the household, as a meso scale unit, challenges dichotomous thinking about pro-environmental practices by drawing together macro and micro level processes. Louise Reid and colleagues (2010) thus call for this multi-scalar reach and inter-scalar potency of the household to be better identified and incorporated into research.

We argue that our application of a material geographies approach to research into household sustainability further develops this call to recognise and conceptualise the household as a crucible of pro-environmental behaviour. A material geographies lens interprets the household as both a social and material entity simultaneously, and provides a useful hook for uncovering the 'missing scale' in contemporary environmental policy debates around urban sustainability. For instance some contributors to this volume have drawn on the insights of embodied geographies and demonstrated that households are (often) social collectives, with different individual members exhibiting and enacting diverse sets of relationships to both the human and non-human networks comprising their dwellings. But perhaps the key example is the way in which various authors have drawn on the idea of hybrid geographies and socio-technical networks as means of moving beyond the assumptions of an atomised decision-making household unit, widening the analysis of sustainable consumption to include broader socio-cultural factors and more complex, materially grounded understandings of agency. By drawing out the context of the household as a domestic unit that operates within broader social networks, the frame of analysis is expanded to consider these networks as sources of influence on household consumption practices. Similarly, rather than treat the individual dwelling as the locus for consumption of material resources, this too must be situated within broader networks of infrastructure and services that support and maintain existing consumption practices. A material geographies approach can thus augment our understanding of how the household mediates between and collates different scales of environmental action.

At the same time this should not be read as a rejection or glossing over of individual agency or the political responsibility that inheres within human–environment interactions. Indeed recognising the household as a hybrid social and material entity draws attention to the importance of human agency in navigating a sustainable way through the complex networks of consumption and disposal that both converge on households and extend well beyond individual dwellings. Collectively, the theorisations articulated in this volume provide a critique of the tendency of Actor Network Theory to flatten out understandings of agency in a manner that can, at times, obscure the hierarchical relationships among the actants (Hughes and Reimer 2004). Instead various authors have developed political analyses that highlight the power relations operating at and between different geographical scales, and the critical role of households in mediating these power relations in ways that advance pro-environmental behaviour. Environmental policies which converse between the national agenda and the individual citizen must normally do so through households. For this reason Reid and colleagues (2010: 323) argue that households are sites of propagation where 'society-wide values are grounded and made practical within the real-world', and thus 'operationalized and shaped, giving rise to and legitimizing new actions'. Through this mediating and propagating function, households draw together multiple networks and multiple scales of action, and become the nexus where processes of environmental governance are transformed into practices of household sustainability, as Aidan Davison surmises in his discussion of the chapters addressing issues of governance and citizenship.

Could this be taken further? Is there a potential for households and householders to be more invested in policies for pro-environmental action? In relation to policies developed by governments at local, state and national levels, householders are constructed not just as rational decision makers, but construed, as Andy Scerri points out, as 'stakeholder citizens'. In this framework households contribute to the social good of 'sustainability' through exercising consumer sovereignty via a privatised ethics of rational choice (Scerri 2006). But given their mediating and propagating function, might households be able to play more of a role in promoting environmental sustainability? Instead of just exercising responsibility by consumption choices, is there scope for households to contribute to the development and management of environmental policies? For instance there might be potential for a co-management approach to policy on environmental sustainability, incorporating communities and governments in decision making. This could be transformative, giving households more power to address policies and processes which affect their consumption patterns. Of course, we are not naive enough to suggest this would necessarily be effective. There are inherent tensions in co-management teams, with different governing actors – community and state – bringing different rationales and motivations.

Key Areas for Further Research

The above paragraph begins to describe some of the potentially productive areas for further research into household sustainability inspired by material geographies. One immediate avenue, then, would be to explore ways to constructively engage stakeholder citizens in environmental decision making at the household scale. This could perhaps entail participatory action research on adaptive co-management between government agencies and community organisations. Since the household is a mediating scale between societal values and individual behaviour, a parallel research direction inspired by the multi-scalar purchase of the household would be to investigate how local actions and initiatives in households could be scaled up to national or regional environmental policies and protocols. Similarly, further research is needed on relational networks both within households and between locally proximate households. Such suggestions are also prompted by Reid and colleagues (2010). Who performs gate-keeping and management roles in terms of learning and propagating sustainable behaviours? How is pro-environmental knowledge disseminated within households and, similarly, how is such knowledge shared between households, such as the practices of sustainable home renovation discussed in the chapter by Ralph Horne, Cecily Maller and Ruth Lane?

The other point to recognise, and which needs to be taken up in future research, is that households are not uniform. They are diverse, differing demographically, geographically and culturally. So while the household scale connects societal attitudes and national policies to individual behaviours, this link is not at all homogenous, and consequently both further complicates the research directions described above and offers another set of research agendas. Demographically there are many household types – not only nuclear families, but couple-only, multi-generational, multi-family, group, single-parent and, the fastest growing household type across much of the West, single- occupancy households. If the household, as a meso scale unit, is an articulation between regional/national change and individual action, then a good next step in understanding and promoting sustainable practices might be to unpack how relational processes inherent in different household compositions vary. And of course households, as we have argued, are also deeply embedded in wider social networks, so understanding household variation around particular practices also means analysing households in different neighbourhood contexts. Comparative work at different scales would be useful here: inner city/

outer suburban, urban/rural and national comparisons, for instance, could yield quite different perspectives on the uptake and applications of pro-environmental behaviours and eco-efficient technologies.

And amid all of this future research we need to keep issues of political economy to the fore. One of the key themes of household sustainability – again demonstrated throughout this volume – is the question of consumption. This in turn raises concerns about the complex interplay of sustainability goals, economic viability and political efficacy. It now seems clear from a range of critiques, including those presented here by Chris Gibson, Gordon Waitt, Lesley Head and Nick Gill, and by Aidan Davison, that environmentally conscious consumption has been co-opted by the market as another means of deriving profit from the circulation of resources and commodities. In short, 'green consumption' remains capitalist consumption, with all the inherent problems associated with growth imperatives. Further useful research, then, could investigate the potential for viable large-scale non-market alternatives for sustainable consumption – alternatives that actually entail less consumption through practices of reuse, and reduce the volume of resources and commodities in market circulation. A number of contributing authors offer such possibilities, for example Horne, Maller and Lane; Watson and Lane; and Dowling and Power. Given the market links between capital accumulation and property ownership, key to this move would be a sophisticated analysis of property relations that extended beyond the formal, tightly bounded property rights associated with markets and offered careful analysis of innovative property arrangements for goods and services. Louise Crabtree's chapter in this volume on up-scaling co-housing begins to develop this agenda, as does work elsewhere on car sharing (Simpson 2009) and alternative food networks (Kneafsey et al. 2007). Our hope is that the material geographies approach we have offered here will inspire and impel new and innovative research into the relational practices and governance of household sustainability.

References

Anderson, B. and Wylie, J. 2009. On geography and materiality. *Environment and Planning A*, 41(2), 318–335.

Hodson, M. and Marvin, S. 2009. 'Urban ecological security': a new urban paradigm? *International Journal of Urban and Regional Research*, 33(1): 193–215.

Hughes, A. and Reimer, S. (eds) 2004. *Geographies of Commodity Chains*. London: Routledge.

Kaika, M. 2004. Interrogating the geographies of the familiar: domesticating nature and constructing the autonomy of the modern home. *International Journal of Urban and Regional Research*, 28(2), 265–286.

Kneafsey, M., Holloway, L., Cox, R., Dowler, E., Venn, L. and Tuomainen, H. 2007. *Reconnecting Consumers, Food and Producers: Exploring 'Alternative' Networks*. Oxford: Berg.

Law, J. 2004. *After Method: Mess in Social Science Research*. New York: Routledge.

Reid, L., Sutton, P. and Hunter, C. 2010. Theorizing the meso level: the household as a crucible of pro-environmental behaviour. *Progress in Human Geography*, 34(3), 309–327.

Scerri, A. (2006) Self-Orienting Individuals: Subjectivity and Contemporary Liberal Individualism. Thesis submitted in partial fulfilment of PhD award, Melbourne, RMIT University.

Simpson, C. 2009. Cars, climates and subjectivity: car sharing and resisting hegemonic automobile culture? *M/C Journal* [Online], 12(4), October 2009. Available at: http://journal.media-culture.org.au/index.php/mcjournal/article/view/176 [accessed: 27 August 2010].

Index

Note: Page numbers with letters, for example, 13f, 56t, indicate figures and tables.